Structures: A Studio Approach

Edmond Saliklis

Structures: A Studio Approach

 Springer

Edmond Saliklis
Department of Architectural Engineering
California Polytechnic State University
San Luis Obispo, CA, USA

ISBN 978-3-030-33155-9   ISBN 978-3-030-33153-5   (eBook)
https://doi.org/10.1007/978-3-030-33153-5

This Springer imprint is published by the registered company Springer Nature Switzerland AG
The registered company address is: Gewerbestrasse 11, 6330 Cham, Switzerland

This book is dedicated to our daughters, Sigita Terese and Ina Marija, and to their husbands, Matthew and William. They are endless sources of joy and love in our lives.

Acknowledgements

Many colleagues at California Polytechnic State University in San Luis Obispo were very helpful to me in creating this suite of materials. Having a supportive community of like-minded peers is one of the joys of being a professor.

In particular, I am extremely grateful to Professor Graham Archer who spent many hours with me discussing the details of this material. He introduced to me the idea of a three-dimensional approach to the Müller-Breslau method. Without the collegial co-teaching of the studios with Professor Ansgar Killing and Professor Clare Olsen, the chapter on shells could not have existed. I am extremely grateful for their willingness to experiment with me in this novel studio setting. Thanks go out to the head of the Architecture Department at Cal Poly, Professor Margot McDonald, and to the head of Cal Poly's Architectural Engineering Department, Professor Allen Estes, who supported our shells studio. Professor Catherine Wetzel of the Illinois Institute of Technology's Architecture Department generously shared her list of precedent buildings with me and guided me on the pedagogy of large deformation physical models in the studio. Professor John Lawson patiently reviewed many of the projects presented here, for that I am very grateful.

Special thanks go out to my student research assistants, Geoffrey Sanhueza, Joshua Lange, Michael Goldenberg, and Paris Allen. This work really could not have been completed without their enthusiastic and able assistance. These research assistants were funded through a generous grant from the William Randolph Hearst Foundation, for which I am grateful.

Contents

Introduction

1

Structural engineering has an intellectually rich and aesthetically pleasing tradition of blending design and analysis. The goal of this book is to reinvent this rich tradition with the hope of nurturing the growth of the next generation of practicing engineers and architects. Today, we have a tremendous opportunity to link design, analysis and computational thinking in a single, cohesive, graphical approach to structures. Today, we can truly integrate structure into architecture, in a way that was intuited by brilliant designers of the past. Sophisticated programmable 3D computer graphics open up a new worldview, one which allows students to see how mathematical knowledge can be used to solve architectural design problems. Linking parametric modeling, with graphical techniques of structural analysis and with prototype model construction, will deeply impact the designers of the next generation.

The structural artist Robert Maillart exemplified "discipline and play" in his work; there were enormous constraints on his designs, but he wilfully chose to celebrate playful aesthetic ideals in his highly disciplined bridges. Similarly, when the engineer Eduardo Torroja described the design process, he was careful to emphasize that it was neither purely rational nor purely imaginative, "but rather both together."

More recently, this quest for "honesty" in structural architecture can be summarized by engineer Jorge Schlaich's sense of public duty as a designer. He feels a strong commitment to create an aesthetically pleasing environment, to have honest, i.e. structural rational forms, to "build well and to build beautifully" ... "ugly structures can destroy the environment and make people sick, they also contribute to the hatred of technology...we engineers cannot push this problem off onto the architects, we must lead by thoughtful design principles", says Schlaich.

One major reason why the structural artist Fazlur Khan was able to realize his dreams was the fact that he thought broadly, artistically and collaboratively. Fazlur Khan epitomized both structural engineering achievement and creative collaborative effort between architect and engineer. Only when architectural design is grounded in structural realities, he believed — thus celebrating architecture's nature as a constructive art, rooted in the earth — can "the resulting aesthetics ... have a transcendental value and quality." Fazlur Khan was constantly inquisitive; thinking, drawing, discussing, and inventing non-stop. He did not shy away from philosophical writings on structures. He believed that skyscrapers embody a distinct sense of focus and purpose and must display clarity in thought and in action. He was deeply concerned with the role of structural logic in new architectural development.

© Springer Nature Switzerland AG 2020
E. Saliklis, *Structures: A Studio Approach*, https://doi.org/10.1007/978-3-030-33153-5_1

Professor David Billington of Princeton University has written eloquently about this concept of "Discipline and Play" in structural design. His teachings have infused this book, and have been reinvented here, with the hope that aesthetically appealing, yet eminently rational designs can spring from the trained, creative minds of the engineer and of the architect. Specifically, the materials in this book hope to blend the training of the engineer and of the architect. A deep understanding of mathematics, of materials, of construction and of good design, allows for an infinite number of shapes that are efficient, economic and elegant. This creative wellspring had its origins in the inventive structural engineering that occurred in the late nineteenth and early twentieth centuries. There, engineers took risks and created structures for which there were no design precedents, much less design codes. They perceived structures as actively and directly responding to externally applied loads. This mindset of engineering was creatively absorbed into the architecture of its time, and most demonstrably, was shown in the creation of tall buildings in nineteenth century Chicago, and in thin shell spatial structures in Italy, Uruguay, Mexico and Switzerland until the 1960s. Structural artists such as Pier Luigi Nervi created magnificent spaces which had sculptural power, yet were extremely rational to design and to construct. The structural artist Felix Candela made incredible thin-shell concrete structures in Mexico City in the mid twentieth century, only because he intelligently blended mathematics, aesthetics and constructability. It is this blending of *Discipline and Play* that readers are encouraged to develop.

The coupling of mathematical rigor, to an engineering intuition of flow of forces in structures, takes time and patience to develop. Even in a polytechnic university such as ours, it is challenging to ensure that the engineer and the architect have a common vocabulary. We believe this is important so that effective integration of the two skill sets can occur in professional practice. In a polytechnic university, one way of linking the skill sets of the two disparate groups, and of attempting to find structural rigor as well as force flow intuition, is through an interdisciplinary studio for architecture and architectural engineering students. These studios have been taught to hundreds of students as an experimental attempt to incorporate structural rationalism into architectural design. It is hoped that the pedagogical tools described in this book, which have arisen from these studios, will have the transformative effect of inspiring both sets of students. The architecture students will have a deeper understanding of the flow of forces in structures, the engineering students will have a greater appreciation of the role of intuition, design and rational form finding. The ultimate goal of this book is to help the reader develop a unique structural sensibility, an ethos that places structural design on an equal footing with the design of program, skin, massing and site. If successful, these exercises will be used at other universities to show students that structural knowledge is fundamental to architectural production. We hope to promote an architectural pedagogy which is inclusive of structure as a form generator, and to do so in a studio setting. When separated from architectural design, structure becomes additive, burdening the student, and the faculty. This book hopes to situate structural principles within the studio, so that structure becomes integral to a way of thinking and that structure becomes an opportunity for design ideas to flourish.

Combining technical and visual literacy, through the exploration of elegant form and the careful study of significant historical precedents will inspire our students and will ignite in them a passion for the blending of technological prowess and aesthetic sensitivity. On a broader level, this book would like to contribute to the goal of ensuring the role of the arts and sciences as a cornerstone of our society, through the nurturing of creative and technical talents.

Admittedly, the balance between technical and visual, or between discipline and play, will lean heavily towards the disciplined rigor that engineering pedagogy is based on. For some readers, the fact that there are equations in this book will be off-putting. For others, the presence of carefully designed posters and the inclusion of sculptural studio models will seem frivolous. As such, readers may tend to jump to chapters of interest, and to totally avoid sections of this book. Could the final chapters on frames and shells be used without a deep exploration of the early chapters on loads, walls and diaphragms? The answer is yes, but the process would then be a hidden "black box" that somehow works. Could quizzes and exams be made from the initial chapters, and the somewhat artistic portions of the later chapters be skipped? Again, the answer is yes, but then the opportunity to blur the lines between engineer and architect would be lost.

In the phrase "discipline and play", discipline comes first. A game is fun, only if there are rules.

Loads

Each component of every building must be designed to safely resist all of the forces that are expected to act on the structure during its lifetime. These forces include Dead Loads (DL) which are vertical (downward) loads due to the weight of all permanent structural and non-structural components of a building such as walls, floors, roofs and fixed service equipment. Yet, one does not truly know the dead load of a building that has not yet been designed! Thus, a good starting point is to use commonly accepted estimates found in building codes. Figure 2.1 shows a sample of such estimates.

kN/m³	kN/m²	lb/ft³	lb/ft²	
	0.3		6	Gravel roofing
	0.1		2.5	1 inch (roughly) wood cladding
	0.05		1	acoustic tile ceiling
	0.2		5	Gypsum ceiling per inch thickness
	0.1		3	plywood per inch thickness
5.5		35		lumber
23.5		150		concrete
18.8		120		brick
				Steel is lb/ft

Fig. 2.1 Commonly accepted dead load estimates

Notice that some of the units are force/area and others are force/volume and steel sections are described with force/length. If Imperial Units are used, note that 1 ft = 12 in.

Live Load (LL) is the load superimposed by the use and occupancy of the building, not including wind load, earthquake load or dead load. It includes the weight of people, furniture etc. Live load varies as people move around or as furniture is rearranged. Although live load is essentially dynamic (i.e. time varying) usually it is assumed to be a static, vertical (downward) load.

The minimum live loads required by the International Building Code (IBC) are found in that Code's Table 1607.1. These loads apply to roof and to floor live loads. Floors in buildings where partition locations are subject to change must be designed to support a load of 15 lb/ft² (0.7 kN/m²) in addition to all other loads.

© Springer Nature Switzerland AG 2020
E. Saliklis, *Structures: A Studio Approach*, https://doi.org/10.1007/978-3-030-33153-5_2

Should we design every structure for every conceivable maximum live load combination? The philosophy behind this idea is that if the loads are not too great (ASCE 4.7.3) and if the area considered is fairly large (ASCE 4.7.2) then a conservative reduction (i.e. not too great) is allowed for Live Load.

The most obvious loads are those due to gravity, induced by self weight, snow and occupancy. Gravity loads that are not part of the permanent (Dead) loads are considered Live Loads. In California, lateral loads are extremely important because of seismicity in our region (most of the California earthquakes excite the building laterally, not vertically, Northridge was an exception). Wind is the enemy of tall buildings, and wind force is proportional to velocity squared (Bernoulli!). Wind speed is faster the higher you go up, inducing rapidly increasing forces for tall buildings.

The *International Building Code* (IBC) has been adopted by many regions. It generally absorbs other codes into it, for example *ASCE 7, **Minimum** Design Loads for Buildings and Other Structures* is a detailed description of loads, and this code has been adopted or absorbed by IBC. Material specific codes have also been adopted by IBC such as the ACI 318 for concrete, and the AISC code for steel. A key idea in all building codes is the word "minimum". It is the legal minimum engineers must abide by.

The weight density of typical structural materials range from very dense steel to very light wood. Note that it is necessary to estimate the self weight of a building before it is actually designed, thus trial sizes are chosen soon in preliminary design because the final design must account for self weight. Thus, certain dead loads (DL) are assumed, which allows for a preliminary design. Then the self weight must be checked to see if the initial assumptions were acceptable or not. Loads are typically expressed as force/length^2 in these calculations. Figure 2.2 shows typical loads that are used in the IBC as preliminary design aids.

Fig. 2.2 Typical structural material self weight

Typical DL	lb/ft²	kN/m²
Roof	15 to 40	0.7 to 1.9
Steel floor	80	3.8
Concrete floor	60	2.8
Light frame wood	10	0.5
Masonry/concrete façade walls	40 to 80	1.9 to 3.8
Partition walls	20	0.9
Stud walls	8 to 12	0.4 to 0.6
Glass curtain walls	12	0.6

A cross sectional drawing of a floor system can be translated into uniform Dead Load (force/length^2) by spreading discrete elements over the plan that they act on. The itemization of all the elements in the cross section, and the subsequent translation of those element loads into a uniform force/length^2 load is done by a so-called "load take-off sheet", essentially an itemization of all the elements acting together on a floor. This technique always focuses on ending up with force/length^2 even if the original values are force/length or force/length^3.

The first step is always to determine the appropriate dead loads for the structural design of a system. The Dead Load is presented in one of three formats: force/length, force/length^2 or force/length^3 for various elements, but we seek some net force/length^2 and then to apply it as a force/length on a given element. This can be done by simply adding up all the weights of the elements and dividing by the floor area acted upon. But another common method in practice is to "widen" element loads that are expressed as force/length, by dividing these loads by the on-center (o.c.) spacing of the discrete elements, thus force/length^2 is obtained.

Figure 2.3 shows seemingly obvious, but necessary, thoughts about multiplying one quantity such as force/length^3 by a length to get force/length^2.

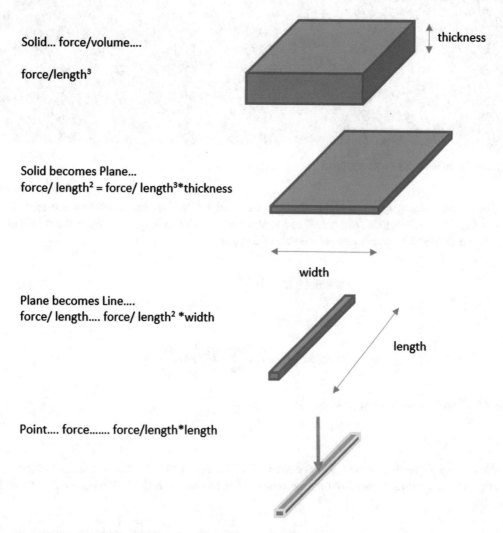

Solid... force/volume....

force/length^3

thickness

Solid becomes Plane...
force/ length^2 = force/ length^3*thickness

width

Plane becomes Line....
force/ length.... force/ length^2 *width

length

Point.... force....... force/length*length

Fig. 2.3 Dimensional analysis volume become line

Consider the elegant Farnsworth House, ostensibly designed by Mies van der Rohe, but really engineered by Myron Goldsmith. Look at the structural logic of this building. First of all, the flow of gravity loads may seem confusing because there are certain elements that are given in force/length^2 such as the ceiling, the roof concrete, etc. Yet other elements such as steel beams which always have a known force/length must be "spread out or expanded" to find force/length^2. Do this through the spacing of the beams on-center (o.c.). A studio approach of looking at the Farnsworth House would be to build a small model of the building. An algebraic approach would begin by studying the roof loads and sequentially moving down the structure to the foundations. Figure 2.4 blends these two approaches.

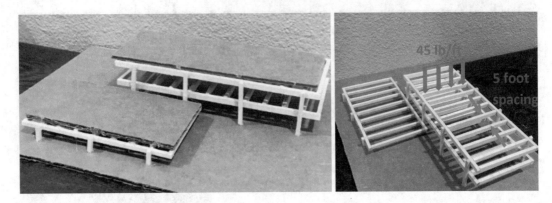

Fig. 2.4 Initial study of Farnsworth House

For instance, using Imperial Units, if each beam weighs 45 lb/ft and the beams are spaced 5ft on-center, the force/length is "expanded" to a force/length^2 as shown in Fig. 2.5. Note, the calculations are shown in MathCad which denotes pounds of force is lbf.

$$\text{W10by45plf} := 45 \, \frac{\text{lbf}}{\text{ft}}$$

$$\text{spacing} := 5\text{ft}$$

$$\text{W10by45psf} := \frac{\text{W10by45plf}}{\text{spacing}} = 9 \, \frac{\text{lbf}}{\text{ft}^2}$$

Fig. 2.5 Changing force/length to force/length^2

Next, suppose these roof beams supported a ***three inch*** concrete slab and ***two inches*** of gypsum ceiling. These are both symbolized by the removeable cardboard in the following model shown in Fig. 2.6.

Fig. 2.6 Dead load on
Farnsworth House model

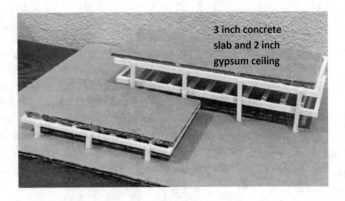

Refer to the building Code, or to the previous tables to find the concrete roof load, and the gypsum ceiling load. These roof loads were chosen not from an extremely practical point of view, but solely to have the experience of changing force/length^3 and force/length^2 per inch worked out once in detail as shown in Fig. 2.7.

$$slabthick := 3\,in \qquad\qquad gypsumthick := 2\,in$$

$$concretepcf := 150\,\frac{lbf}{ft^3} \qquad gypsumwt := \frac{5\,\dfrac{lbf}{ft^2}}{in}$$

$$concretepsf := concretepcf \cdot slabthick = 37.5\,\frac{lbf}{ft^2}$$

$$gypsumpsf := gypsumthick \cdot gypsumwt = 10\,\frac{lbf}{ft^2}$$

Fig. 2.7 Ensuring all dead load is force/length^2

So the "load take-off" is the summation of the beam weight, the concrete slab weight, and the gypsum weight, all expressed as force/length^2.

$$floorload = W10by45psf + concretepsf + gypsumpsf = 56.5\,lb/ft^2 \qquad (2.1)$$

A few final notes on "load take-offs". The Mechanical, Electrical and Plumbing (MEP) loads vary greatly over any building, in some spots this load will be large, in other spots this load is non-existent. It is typical practice to simply add some MEP weight uniformly throughout to account for this load. MEP is often 4 lb/ft^2 or 5 lb/ft^2 (0.19 kN/m^2 to 0.24 kN/m^2).

There is another category called Miscellaneous (MISC). This is very useful in that it is a little extra load that you can reduce later if your beam sizes increase during design, think of it as a secret safety fund that you can draw from when need be.

Before leaving the Farnsworth House, consider the Lateral Force Resisting System (LFRS). Is it there? If so, where? Myron Goldsmith cleverly hid the LFRS along gridlines that contain pairs of columns. Those bays are slightly different in their connections, they form moment resisting frames, an important LFRS. Moment resisting frames allow the space to be open and airy, but still provide lateral resistance. However, they are the most expensive LFRS and the least efficient. Figure 2.8 shows the structural logic of the LFRS.

Fig. 2.8 Lateral force resisting system in Farnsworth House model

One more example is described in Fig. 2.9. A typical cross section through a floor is shown. There is a concrete slab monolithic to a series of concrete beams 8ft on-center (o.c.). Assume this pattern of T Beams repeats every 8 feet indefinitely. There is MEP and there is a ½ inch Gypsum ceiling. Find a net force/length^2 on this floor.

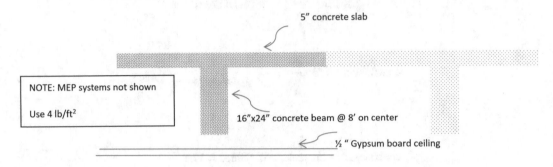

Fig. 2.9 Cross section for load takedown example

The solution is to create a so-called "load takeoff" sheet. A typical calculation is shown in Fig. 2.10.

Fig. 2.10 Load takedown calculations

$$\text{MEP} := 4\,\frac{\text{lbf}}{\text{ft}^2}$$

$$\text{slabthick} := 5\,\text{in} \qquad \text{gypsumthick} := 0.5\,\text{in}$$

$$\text{concretepcf} := 150\,\frac{\text{lbf}}{\text{ft}^3} \qquad \text{gypsumwt} := \frac{5\,\frac{\text{lbf}}{\text{ft}^2}}{\text{in}}$$

$$\text{concreteslabpsf} := \text{concretepcf} \cdot \text{slabthick} = 62.5\,\frac{\text{lbf}}{\text{ft}^2}$$

$$\text{gypsumpsf} := \text{gypsumthick} \cdot \text{gypsumwt} = 2.5\,\frac{\text{lbf}}{\text{ft}^2}$$

$$\text{beampsf} := \frac{[16\text{in} \cdot (24\text{in} - 5\text{in})] \cdot \text{concretepcf}}{8\text{ft}} = 39.583\,\frac{\text{lbf}}{\text{ft}^2}$$

notice that here is length*length* force/length³ divided by length

Sum up all the components for the final force/length^2 value.

$$floorpsf = MEP + concreteslabpsf + gypsumpsf + beampsf = 108.6\,{}^{lb}\!/_{ft^2} \qquad (2.2)$$

The next example uses the same idea of "expanding" discrete beam weights (force/length) into force/length^2 through the on-center (o.c.) spacing. This will be version 1 of some steps in the following calculations. Version 2 will simply add up the weights of the elements and dividing that total weight by the area they act on.

The layout of the example problem is in Fig. 2.11. The spacing of Gridlines A, B and C are 40ft. The spacing of Gridlines 1,2 and 3 are 30ft. The heights of the columns are 12ft but that information will

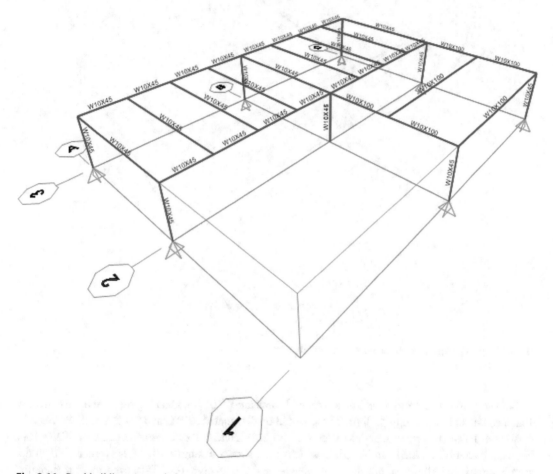

Fig. 2.11 Steel building geometric layout

not be used in this example

In practice groups of beams, groups of girders and groups of columns are labeled with a generic designation. They are not specifically named in structural drawings as W10×45, such information would be provided in a beam schedule. In this example, it would be more efficient to say B1 is a W10×45 and B2 is a W10×100. It is reasonable to label the elements along Gridlines 2 and 3 as girders, not beams, but such a distinction need not be used here. Beams frame into girders, so beams have a lower hierarchical standing than do girders. Yet in this example, labels will be shown only for B1 (the W10×45s) and B2 (the W10×100s). The two beam groups are clearly labeled in Fig. 2.12.

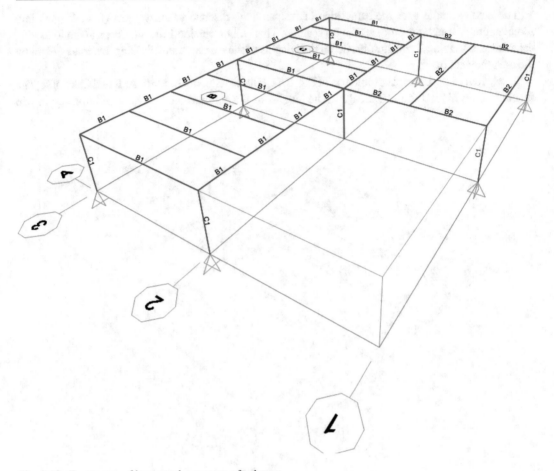

Fig. 2.12 Two groups of beams and one group of columns

Between Gridlines 2 and 3, all the horizontal beams are W10×45, meaning they are approximately 10inches tall and they weigh 45 lb/ft. These would be designated as B1 on drawings and described in detail as W10×45 in a beam schedule. Between Gridlines 1 and 2, the horizontal beams are W10×100, these are called B2 on the beam schedule which is not shown. Using W10×100s keeps the depth the same approximate 10inches, but these are heavier and they weigh 100 lbf/ft. Find the force/length^2 acting on these two distinct areas. Consider both:

- **VERSION 1**: the on-center (o.c.) spacing method
- **VERSION 2**: adding up the weights manually and dividing by the area they act on

Do not double count the beams on Gridline 2 that are common to both areas. Ignore the weight of the columns in this exercise, the floor load is sought. Typical calculations are layed out in Fig. 2.13.

$$\text{spacingAB} := 40\text{ft} \qquad \text{spacingBC} := 40\text{ft} \qquad \text{spacing12} := 30\text{ft} \qquad \text{spacing23} := 30\text{ft}$$

$$\text{plf45} := 45\frac{\text{lbf}}{\text{ft}} \qquad\qquad \text{plf100} := 100\frac{\text{lbf}}{\text{ft}}$$

$$\text{spacingoc} := 10\text{ft}$$

$$\text{psfBeamsABC} := \frac{\text{plf45}}{\text{spacingoc}} = 4.5\frac{\text{lbf}}{\text{ft}^2}$$

$$\text{psfBeamsABCv2} := \frac{9 \cdot \text{plf45} \cdot \text{spacing23}}{(\text{spacingAB} + \text{spacingBC}) \cdot \text{spacing23}} = 5.062\frac{\text{lbf}}{\text{ft}^2}$$

$$\text{psfBeams23} := \frac{\text{plf45}}{0.5 \cdot \text{spacing23}} = 3\frac{\text{lbf}}{\text{ft}^2}$$

Fig. 2.13 Initial load takedown of steel beams

First focus on the nine parallel W10×45 beams spanning between Gridlines 2 and 3. Notice that simply "widening" the dead load of the beams from force/length to force/length^2 through the operation of dividing by the on-center spacing, is not exactly the same as adding up all nine beam weights and dividing by the area they act on. But from an individual beam's point of view, the "widening" of load make sense, this is the tributary width that affects any one interior beam. The problem is that the beams on the extreme edges are also assumed to act on the same areas as the interior beams, thus, the technique slightly overstimates the area and consequently understimates the force/length^2 arising from dead load of the beams.

Why is this Version 1 useful? The answer is that this method is ideally suited for exploratory, initial design ideas. The building is not designed yet! Laying out a possible beam spacing allows for quick calculations of floor loads. Over time and an accumulation of design experience, the patterns of typical floor loads can be remembered. For example, 4 or 5 lb/ft^2 is a typical floor load in some situations, depending on the material being used. Timber is the lightest structural material we use.

Another advantage of the Version 1 method is that it allows for the establishment of a hierarchy of loads. What that means is that, in any example similar to the frame being studied, the:

- Beams will have the highest lb/ft^2 load, perhaps 4 lb/ft^2
- The girders will have the second lb/ft^2 load, perhaps 3 lb/ft^2, but the area is significantly larger than the area loading the beams

To reiterate what Version 1 of this method is, recall the values obtained. These are highlighted in Fig. 2.14.

Fig. 2.14 Checking force/
length^2 two different ways

$$\text{psfBeamsABC} := \frac{\text{plf45}}{\text{spacingoc}} = 4.5 \cdot \frac{\text{lbf}}{\text{ft}^2}$$

$$\text{psfBeamsABCv2} := \frac{9 \cdot \text{plf45} \cdot \text{spacing23}}{(\text{spacingAB} + \text{spacingBC}) \cdot \text{spacing23}} = 5.062 \frac{\text{lbf}}{\text{ft}^2}$$

$$\text{psfBeamsABCalternate} := \frac{\text{plf45}}{\left(\dfrac{80\text{ft}}{9}\right)} = 5.062 \frac{\text{lbf}}{\text{ft}^2}$$

Version 1 simply uses the on-center spacing. Version 2 sums up all the load and divides by the area that this load acts on. Students need to be comfortable with the on-center spacing estimate technique (Version 1) as well as the summative, precise method (Version 2). In this example both techniques are used.

Next look at the beams along Gridlines 2 and 3. Both techniques were used to find the net force/length^2 of these beams. The first technique "widened" the load by the tributary width, the second technique added up all the weights and divided by the area they act on. Both techniques, shown in Fig. 2.15, gave the same final answer.

Fig. 2.15 Checking force/
length^2 twice in new area

$$\text{psfBeams23} := \frac{\text{plf45}}{0.5 \cdot \text{spacing23}} = 3 \frac{\text{lbf}}{\text{ft}^2}$$

$$\text{psfBeams23v2} := \frac{\text{plf45} \cdot (\text{spacingAB} + \text{spacingBC}) \cdot 2}{(\text{spacingAB} + \text{spacingBC}) \cdot \text{spacing23}} = 3 \frac{\text{lbf}}{\text{ft}^2}$$

Notice here the hierarchy of loads described previously. The beams parallel to Gridlines A, B and C experience roughly 5 lb/ft^2. The girders along Gridlines 2 and 3 experience 3 lb/ft^2. The final load in this area between Gridlines 2 and 3 is the superposition of the two loads.

$$psfArea23 = psfBeamsABC + psfBeams23 = 7.5 \; {}^{lb}/_{ft^2} \tag{2.3}$$

The more accurate technique of finding the loads (version2) gives a slightly different value.

$$psfArea23v2 = psfBeamsABCv2 + psfBeams23 = 8.06 \; {}^{lb}/_{ft^2} \tag{2.4}$$

For the heavily weighted portion between Gridlines 1 and 2, since there are so few elements it is easiest to use only the second technique, namely to add up the weights and divide by the area. Note that beams on Gridline2 are common to both areas and are not double counted.

$$psfArea12 = \frac{2\,plf\,100\,spacing12 + 2\,plf\,100\,spacingBC}{spacing12\;spacingBC} = 11.67 \; {}^{lb}/_{ft^2} \tag{2.5}$$

Obtaining the final dead load is quick, now that each force/length^2 in the two distinct areas is known. Calculations are in Fig. 2.16.

Fig. 2.16 Final dead load version 1

$$\text{Load1} := \text{psfArea23} \cdot (\text{spacingAB} + \text{spacingBC}) \cdot \text{spacing23} = 18000\,\text{lbf}$$

$$\text{Load2} := \text{psfArea12} \cdot (\text{spacingBC} \cdot \text{spacing12}) = 14000\,\text{lbf}$$

$$\text{TotLoad} := \text{Load1} + \text{Load2} = 32000\,\text{lbf}$$

A slightly more precise answer can be found using the second version of the load calculation for the 9 beams. Figure 2.17 shows this answer.

Fig. 2.17 Final dead load version 2

$$\text{Load1v2} := \text{psfArea23v2} \cdot (\text{spacingAB} + \text{spacingBC}) \cdot \text{spacing23} = 19350\,\text{lbf}$$

$$\text{Load2} := \text{psfArea12} \cdot (\text{spacingBC} \cdot \text{spacing12}) = 14000\,\text{lbf}$$

$$\text{TotLoad} := \text{Load1v2} + \text{Load2} = 33350\,\text{lbf}$$

It is instructive to qualitatively see how the frame bends due these loads. Figure 2.18 shows qualitative bending moments acting on statically indeterminate elements, which have moment carrying connections at their ends. The beams on Gridline2 have significantly larger moments than do any other elements.

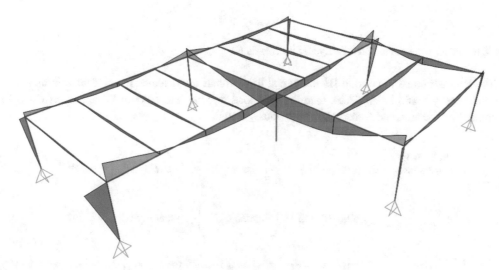

Fig. 2.18 Qualitative bending moment diagrams for steel building

The column loads have not yet been addressed. Suppose the axial load in Column B3 in Figs. 2.11 or 2.12 was sought. There are several ways of finding this axial load:

- Intuiting which areas affect said column, then multiplying the force/length^2 by this area
- Load tracing, element by element and eventually summing up all the shears framing into this column
- The Müller Breslau method

The second method is lengthy and tedious. It can be used as a laborious check. In this example, the first and the third method will be described. These are both considered to be "load tracing" whereas up to now the process has been to define the loads themselves. But both methods of tracing load will be extensively studied in subsequent chapters, so this is a good transition as the last point in this chapter on loads.

Version 1 Intuitive Process The process is highlighted in Fig. 2.19.

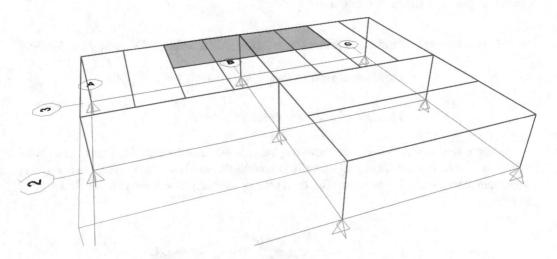

Fig. 2.19 Intuitive process to find tributary area for column B3

The load attributed to column B3 is intuited to span half the distance to adjacent columns. These areas are easily found in Fig. 2.20, then the total load flowing into column B3 is that area (length^2) multiplied by the assigned force/length^2 previously found.

Fig. 2.20 Calculation for column B3 using intuitive process

$$\text{ColB3areapart1} := \left(\frac{1}{2} \cdot \text{spacing23} \right) \cdot \left(\frac{1}{2} \text{spacingAB} \right) = 300 \text{ft}^2$$

$$\text{ColB3areapart2} := \left(\frac{1}{2} \cdot \text{spacing23} \right) \cdot \left(\frac{1}{2} \text{spacingBC} \right) = 300 \text{ft}^2$$

$$\text{ColB3force} := (\text{ColB3areapart1} + \text{ColB3areapart2}) \cdot \text{psfArea23} = 4500 \text{lbf}$$

Version 3 Müller Breslau Method The Müller Breslau Method will be studied in greater detail later. For now, realize it is an energy method. The energy of one element, such as column B3, moving through some known distance such as 1ft, must equal the net forces moving through the related "lofted" distances induced by the lift of the column.

The method becomes incredibly simple because complicated loads, and openings in floors and changing of direction of beams all fall away, simply find the centroid of each part of the force/length^2 that is lofted, and find out how much it lofts, usually ½ of ½ of 1ft. These values are shown in Fig. 2.21

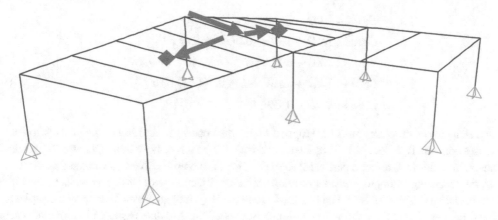

Fig. 2.21 Müller-Breslau geometry for column B3 force

The force in the column is found through the extremely brief steps shown in Fig. 2.22.

Fig. 2.22 Calculation for column B3 using Müller-Breslau

$$\text{ColumnForce} \cdot 1\text{ft} = \text{forcei} \cdot \text{lofti} + \text{forcej} \cdot \text{loftj} + ..$$

$$\text{Areai} := \text{spacingAB} \cdot \text{spacing23} = 1200\text{ft}^2$$

$$\text{Areaj} := \text{spacingBC} \cdot \text{spacing23} = 1200\text{ft}^2$$

$$\text{forcei} := \text{Areai} \cdot \text{psfArea23} = 9000\text{lbf}$$

$$\text{forcej} := \text{Areaj} \cdot \text{psfArea23} = 9000\text{lbf}$$

$$\text{ColB3MB} := \text{forcei} \cdot 0.25 + \text{forcej} \cdot 0.25 = 4500\text{lbf}$$

Notice that each complete area, on either side of the "fold" must be found, and its centroid must be found, and the amount it lofts must be found. The loft was 1ft, go down one slope halfway (½), go down other slope halfway (½ of ½ = ¼). These movements were shown in Fig. 2.21.

The topic of load combinations closes Chap. 2. The most interesting feature of these combinations is that the effect of individual loads can be parametrically studied. Begin with Sect. 2.3 of ASCE 7 which lists the required scenarios that must be studied when considering load combinations. ASCE 7 calls dead load D and live load L in these equations, with roof live load being Lr, Snow is S and Rain is R, Wind is W.

$$Case\,1:1.4D$$
$$Case\,2:1.2D+1.6L+0.5(Lr\ or\ S\ or\ R)$$
$$Case\,3:1.2D+1.6(Lr\ or\ S\ or\ R)+(L\ or\ 0.5W) \qquad (2.6)$$
$$Case\,4:1.2D+1.0W+L+0.5(Lr\ or\ S\ or\ R)$$
$$Case\,5:0.9D+1.0W$$

All elements of structures must be designed so that they can withstand any of the load combinations described in ASCE 7 Sect. 2.3. It is best to consider effects, i.e. how does DL affect a particular element, how does LL affect a particular element. Then, scale up the loads to worsen the effect.

In the following example, consider only one building direction with two shear walls as the Lateral Force Resisting System (LFRS). The DL, the LL and the W (wind) are given, but the W can strike from either direction. Figure 2.23 shows the building, but notice the grid data is shown in terms of spacing, not in terms of Cartesian Coordinates. In Fig. 2.23, consider the first diaphragm a floor and the second diaphragm a roof. Recall from ASCE 7 that roof live loads are handled independently of other live loads.

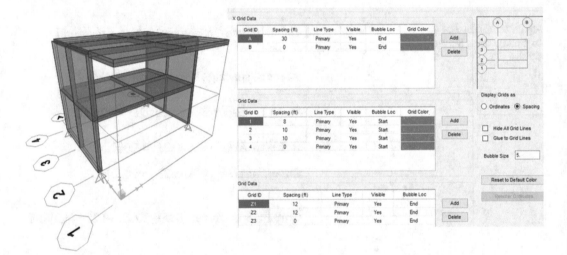

Fig. 2.23 Geometry of two story building load combination example

In Fig. 2.23, notice that the shear walls are pinned at the far base corners and are free everywhere else. What is meant by "study the effect" of a load? Suppose the wind load W was known and it was 12,000lb hitting each diaphragm in the − y direction. This is shown in Fig. 2.24.

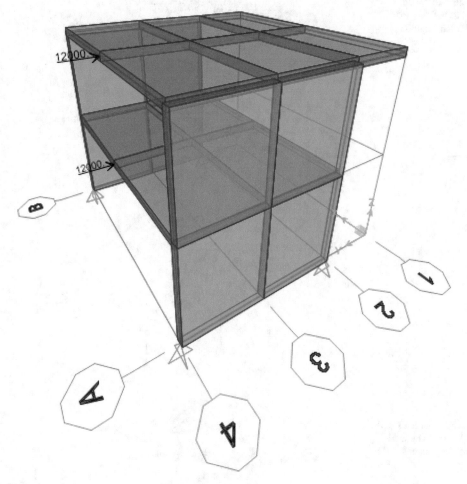

Fig. 2.24 Lateral loads applied to each diaphragm in −Y direction

Also suppose that DL for the floor was 100lb/ft^2 and Lr for the roof was 40lb/ft^2. How might these three loads individually affect the vertical force at A4? The roof load is shown in Fig. 2.25 and the floor load is shown in Fig. 2.26.

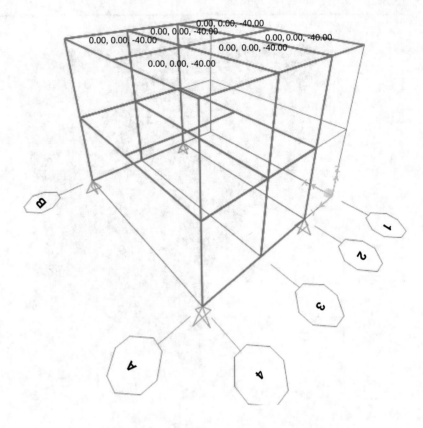

Fig. 2.25 Roof load applied to building load combination example

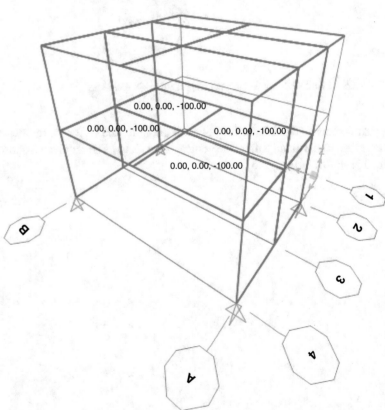

Fig. 2.26 Floor load applied to building load combination example

Project 2–1 The following is a project that can be modified easily. Studying this project and re-creating the values obtained will help develop computational thinking, i.e. the ability to create a model that is nimble enough to respond to changes, but simple enough to clearly reflect hand calculations.

For the two story structure shown above, find the force in the reaction at column A4 for the five load cases described in ASCE 7, Sect. 2.3.1. Do this by hand. Then try to do this via a programming environment. Any language will suffice.

Here are some hints about how to do this, by studying the effect on the reaction at A4 due to each load pattern individually. First, study the effect of DL only, do not apply any other loads.

Figure 2.27 shows some elementary parametric modeling steps. Worth noting in detail are the two versions of finding the reaction at A4 due to DL (RA4DL). There is an inflexible way of doing this, which is not recommended, and a flexible way of finding this, which is preferred. The inflexible manner simply takes the weight due to dead load and divides by 4, since there are two pins on each of the two shear walls. The more flexible way is to describe RA4DL as a function of any load, be it 1.0DL or 1.4DL. Both are shown in the calculations of Fig. 2.27.

$$LAB = 30ft$$

$$L12 = 8ft \quad L23 = 10ft \quad L34 = 10ft$$

$$Z12 = 12ft \quad Z23 = 12ft$$

$$DL = 100\frac{lb}{ft^2} \quad RoofLoad = 40\frac{lb}{ft^2} \quad W = 12000lb$$

$$WeighDL = DL \cdot (L23 + L34) \cdot LAB = 60000lb$$

$$RA4DLinflexible = \frac{WeightDL}{4} = 15000 \; lb$$

$$RA4DL(DL) = \frac{DL \cdot (L23 + L34) \cdot LAB}{4}$$

$$RA4DL\left(100\frac{lb}{ft^2}\right) = 15000lb$$

Fig. 2.27 Typical initial steps for load combination parametric study project

A finite element structural analysis program such as SAP2000 could be used to verify the hand calculations. Typical output is shown in Fig. 2.28.

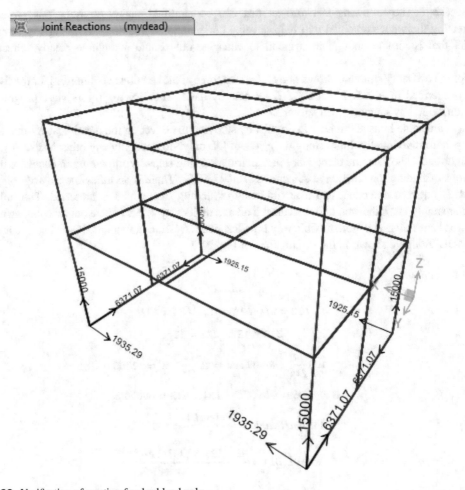

Joint Reactions   (mydead)

Fig. 2.28 Verification of reaction for dead load only

The nimbleness of the parametric approach is shown by simply using some factor on DL to a properly programmed equation.

$$RA4DL(1.4\ DL) = 21000\ lb \qquad (2.7)$$

The value of 21,000lb is verified using SAP2000 which also allows for scaling of load patterns. Figure 2.29 provides validation of this scaled up value.

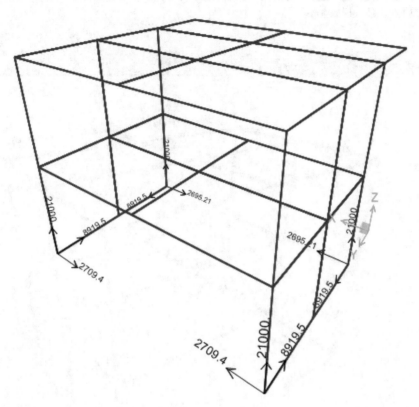

Fig. 2.29 Verification of reaction for Case 1 load combination

Now study the effect of the Roof Load separately. This means, apply only the roof load. Do this in the most flexible way possible, such that you can later scale up or scale down your Roof Load. In Fig. 2.30, an algebraic moment equation was used.

$$Sum\ moments\ about\ line\ 2\ for\ half\ the\ building\ to\ find\ RA4roof$$

$$RA4Roof(RoofLoad) = \frac{\frac{LAB}{2} \cdot (L12 + L23 + L34) \cdot RoofLoad\left(\frac{L12 + L23 + L34}{2} - L12\right)}{L23 + L34}$$

$$RA4Roof(RoofLoad) = 5040\ lb$$

$$RA4Roof(1.6 \cdot RoofLoad) = 8064\ lb$$

Fig. 2.30 Asymmetry of roof load requires moment equation

Notice how flexible this is, the roof load can be applied with any desired factor and its effect on the reaction at A4 is not affected by the DL. This is what is meant by studying each effect separately. The verification of the effect of the roof load is shown in Fig. 2.31 for an unscaled Roof Load and Case 2 of the ASCE Load Combinations.

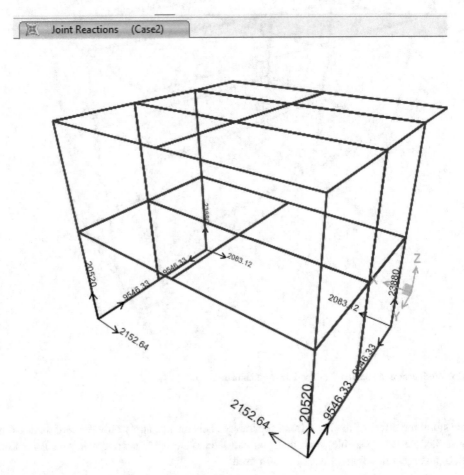

Fig. 2.31   Verification of reaction for Case 2 load combination

Figure 2.32 summarizes the reaction at A4, as it is induced by DL alone, by RoofLoad alone and by Wind alone, then the parametric study allows for the scaled up effects of each of these in the five combinations listed by ASCE 7.

Project 2–2   The following is another exercise in load combinations. It can be solved by hand or via a simple programming environment such as Excel, MathCad, MATLAB, GeoGebra, etc.

A single, two-dimensional frame is subjected to three load patterns. Spacing between A and B is 20ft. Story heights are 12ft. Find the vertical reaction at the base of the frame on gridline B, as highlighted in Fig. 2.32.

$$RA4DL(DL) = 15000\,lb$$

$$RA4Roof(RoofLoad) = 5040\,lb$$

$$RA4Wind = 10800\,lb$$

$$RA4case1 = RA4DL(1.4 \cdot DL) = 21000\,lb$$

$$RA4case2 = RA4DL(1.2 \cdot DL) + RA4Roof(0.5 \cdot RoofLoad) = 20520\,lb$$

$$RA4case3 = RA4DL(1.2 \cdot DL) + RA4Roof(1.6 \cdot RoofLoad) + 0.5 \cdot (RA4Wind) = 31464\,lb$$

$$RA4case4 = RA4DL(1.2 \cdot DL) + 1 \cdot RA4Wind + RA4Roof(0.5 \cdot RoofLoad) = 31320\,lb$$

$$RA4case5 = RA4DL(0.9 \cdot DL) + 1 \cdot RA4Wind = 24300\,lb$$

Case 3 is the most extreme load.

Fig. 2.32 Hand calculations for five load case combinations Project 2–2

Lateral Wind Loads (W) are 1000lb, 1200lb, 1400lb and 1600lb as shown in Fig. 2.33, applied to each diaphragm. Wind load only comes from the west as shown.

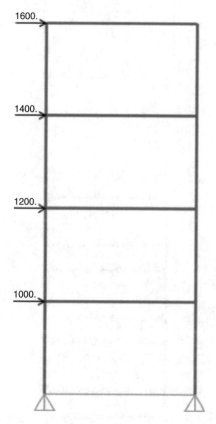

Fig. 2.33 Lateral loads for Project 2–2

DL is 150lb/ft as shown in Fig. 2.34 LL is 70lb/ft applied vertically to each floor as shown in Fig. 2.35. Use the first five ASCE Load Combinations to find the worst possible vertical base support reaction at B. No roof loads are applied, all are LL.

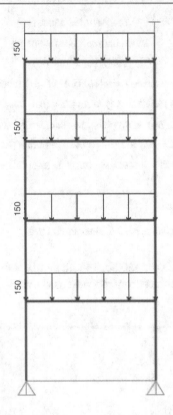

Fig. 2.34 Dead load for Project 2–2

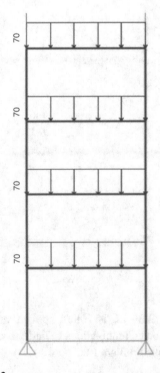

Fig. 2.35 Floor live load for Project 2–2

Dead Load Flow

3

The study of load tracing begins with gravity loads and a grid of horizontal elements. The primary elements are called girders, the secondary elements are called beams. Secondary implies that these members have a lower hierarchical standing, they rely on (connect to) the primary members which we call girders. The beams may be long, or they may be short. Architectural concerns may call for having longer beams connecting to girders, which may have shorter spans but subsequently larger bending moments. Good practice would call for having both of these elements at roughly the same depth so that floor thickness remains somewhat constant. In timber and steel construction, it is common to assume that the secondary beams are simply supported at their ends, i.e. they do not transfer bending moment to the girders. This is symbolized by a gap in the drawing of the connection. In this chapter, the girders are also always assumed to be simply supported to the columns, but along their length they carry moment across each connection. Thus, in Fig. 3.1, the girder passing from span AB to span BC does transfer moment at that intersection, whereas the beam along B connects simply to the girder at B, no moment is transferred.

First consider a grid that is not perfectly symmetric. Figure 3.1 shows this example. Notice the gaps between elements. In structural drawings, such gaps represent connections that do not transfer bending moment. These are known as "simple" connections. Here, the long girder is simply supported to the columns.

© Springer Nature Switzerland AG 2020
E. Saliklis, *Structures: A Studio Approach*, https://doi.org/10.1007/978-3-030-33153-5_3

Fig. 3.1 Geometry of
asymmetric grid example

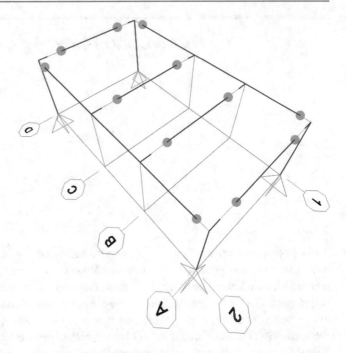

Assume the shorter beams carry a floor load. The floors load the short beams, and the beams transfer load to the long girders. In this model, the floor does not load up the long girders directly, the floor indirectly loads the girders through the load path of the beams. This is known as one-way bending. One way bending is symbolized in Fig. 3.2 with the black, double-headed arrows.

Fig. 3.2 One way bending
direction

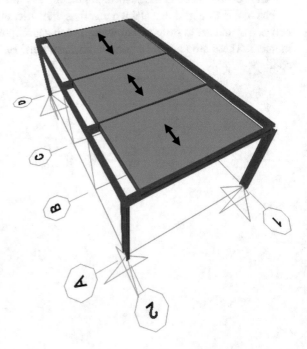

The task at hand is to change a uniformly distributed area load (force/length^2) into a uniform line load (force/length). This is easiest to imagine via a tributary width of each beam. For example, the beam on gridline C experiences half of the load between lines B and C and half of the load between lines C and D. Each beam will experience a uniform load as force/length. Since each beam is simply supported at the point where it connects to the long girders, the force flowing into the girders is ½ the force on each small beam. The girder then feels a series of point loads (force), not force/length, not force/length^2. These calculations are in Fig. 3.3.

Fig. 3.3 Changing force/length^2 to force/length

$$\text{span12} := 20\text{ft}$$

$$\text{spanAB} := 10\text{ft} \qquad \text{spanBC} := 10\text{ft} \qquad \text{spanCD} := 16\text{ft}$$

$$\text{psf} := 110\,\frac{\text{lbf}}{\text{ft}^2}$$

$$\text{plfA} := \text{psf} \cdot \frac{1}{2} \cdot \text{spanAB} = 550\,\frac{\text{lbf}}{\text{ft}}$$

$$\text{plfB} := \text{psf} \cdot \left(\frac{1}{2} \cdot \text{spanAB} + \frac{1}{2} \cdot \text{spanBC} \right) = 1100\,\frac{\text{lbf}}{\text{ft}}$$

$$\text{plfC} := \text{psf} \cdot \left(\frac{1}{2} \cdot \text{spanBC} + \frac{1}{2} \cdot \text{spanCD} \right) = 1430\,\frac{\text{lbf}}{\text{ft}}$$

$$\text{plfD} := \text{psf} \cdot \frac{1}{2} \cdot \text{spanCD} = 880\,\frac{\text{lbf}}{\text{ft}}$$

The force/length images are in Fig. 3.4.

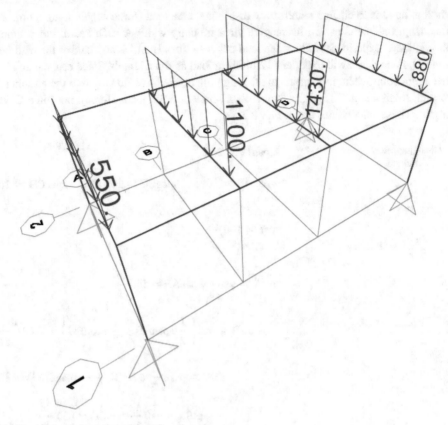

Fig. 3.4 Graphical representation of force/length on beams

Each of the short beams will bend into a circular arc segment shape, due to the uniformly applied load and simple (non-moment transferring) connections to the large girders. Each of the peak bending moments will be $wL^2/8$ for these beams, but each w on a particular beam is based on its tributary width. The girders will have changes in curvature in their deformed positions because they experience point loads, not uniform loads. The greatly exaggerated deformation is in Fig. 3.5.

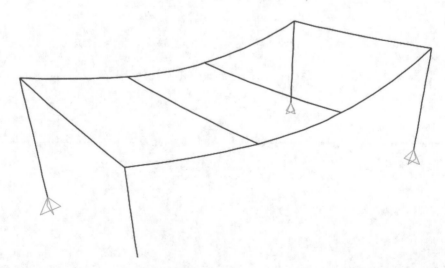

Fig. 3.5 Exaggerated deformation of entire structure

Notice in Fig. 3.6 how the bending moment diagrams are parabolic for the beams ($wL^2/8$) but are linear for the girders.

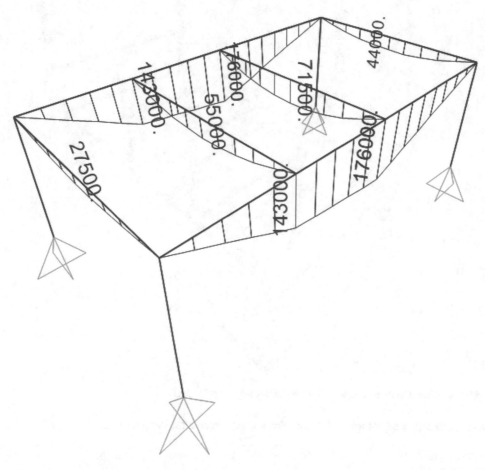

Fig. 3.6 Bending moment diagrams of entire structure

What will the peak bending moment be for the girders? Will it be $wL^2/8$? Fig. 3.7 clearly proves that the answer is "no".

$$\text{pointA} := \frac{\text{plfA} \cdot \text{span12}}{2} = 5500 \, \text{lbf}$$

$$\text{pointB} := \frac{\text{plfB} \cdot \text{span12}}{2} = 11000 \, \text{lbf}$$

$$\text{pointC} := \frac{\text{plfC} \cdot \text{span12}}{2} = 14300 \, \text{lbf}$$

$$\text{pointD} := \frac{\text{plfD} \cdot \text{span12}}{2} = 8800 \, \text{lbf}$$

Fig. 3.7 Calculating the point load magnitudes

The flow of these loads is graphically presented in Fig. 3.8.

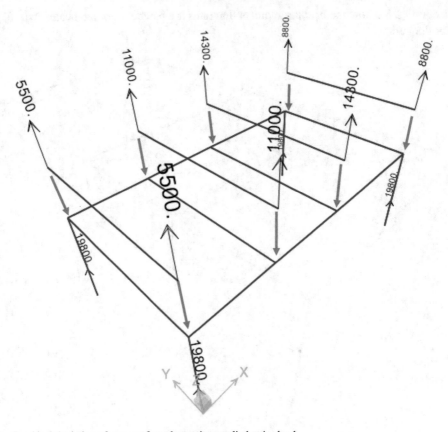

Fig. 3.8 Graphical depiction of support force becoming applied point load

These upwards support forces become downward forces onto the girders as shown in Fig. 3.9.

Fig. 3.9 Graphical
depiction of only the
applied point loads

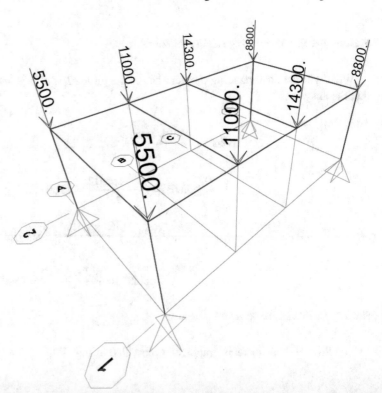

Figure 3.10 shows the resulting bending moment diagrams.

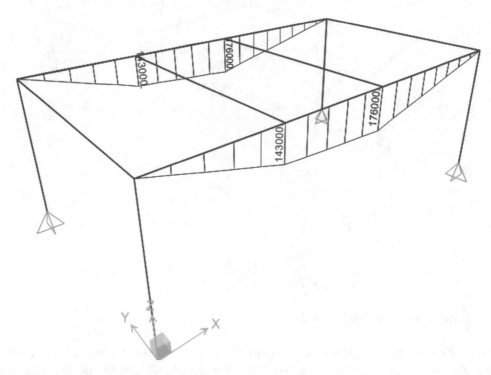

Fig. 3.10 Bending moment diagrams on girders are not parabolic

Notice this sequential technique of applying the upward reactions supporting the beams as downward point loads applied to the girders.

To prove the static equivalence of the original area floor load (force/length^2), versus the uniform load on the beams (force/length) versus the point loads on the girders (force), consider the load on the columns. First look the original area floor load, as shown in Fig. 3.11. By symmetry it is clear that each column will carry one-fourth of the total load.

Fig. 3.11 Intuitive approach to column force

$$\text{totalArea} := \text{span12} \cdot (\text{spanAB} + \text{spanBC} + \text{spanCD}) = 720\,\text{ft}^2$$

$$\text{totalForce} := \text{psf} \cdot \text{totalArea} = 79200\,\text{lbf}$$

$$\text{columnForce} := \frac{\text{totalForce}}{4} = 19800\,\text{lbf}$$

The axial load in the columns is graphically represented in Fig. 3.12.

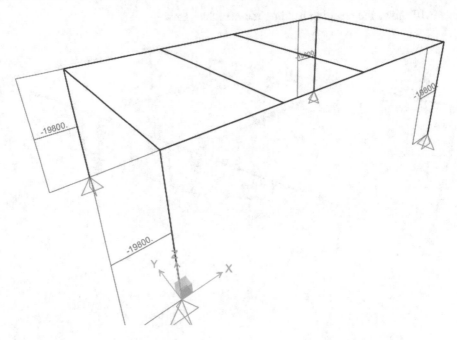

Fig. 3.12 Graphical depiction of column force

It is useful to see how the entire structural system deforms. The power of current structural analysis software allows for such visualizations. Hopefully, they can provide a better intuition of how load flows through the structure. Figure 3.13 is one such image, it is greatly exaggerated but drawn to scale.

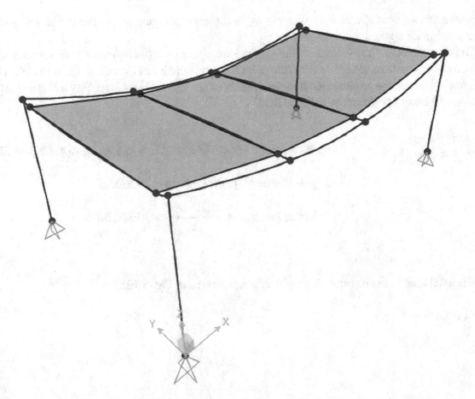

Fig. 3.13 Exaggerated deformation of entire structure including one way bent areas

One single column can be analyzed independently of any beam or girder elements. It is intuitive to calculate how much area a single column is responsible for, or in other words what tributary area is assigned to it. Columns are labeled by two gridline indices, thus the column at the origin is called A1. It is responsible for an area half the distance between 1 and 2, and half the distance between A and D. This intuitive approach to assigning tributary area to a column is shown in Fig. 3.14.

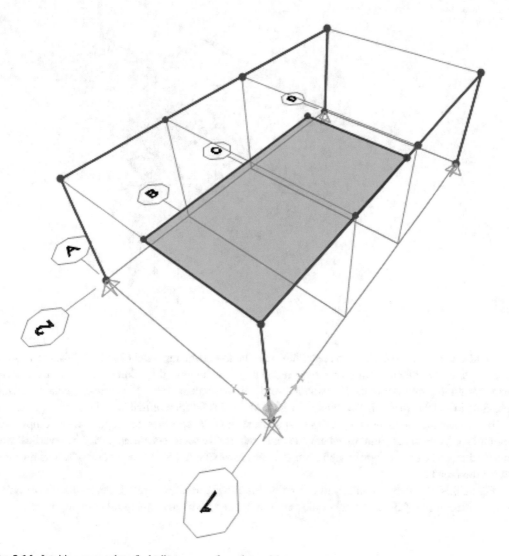

Fig. 3.14 Intuitive approach to find tributary area for column A1

All of the force coming from the assigned tributary area for A1 goes vertically into the column at A1, i.e. one-fourth of the floor load as before.

Yet another way of looking at the column axial force is to show the beam reaction from Beam A and the Girder reaction from Girder 1 connecting to the column at A1. The axial force in the column must be the sum of these two connection forces. Here, the reaction of the beam on Gridline 1 is 5500lb and the reaction of the girder on Gridline A is 14,300 lb. These both connect to the column on A1 thus the axial force in A1 is 19,800lb as before. An exploded view of these forces is shown in Fig. 3.15, with 5500lb + 14,300lb equaling 19,800lb.

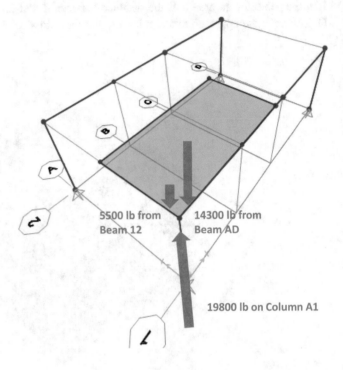

Fig. 3.15 Checking force on column A1 by means of two connections

Load tracing analysis, via statics, starting with the lowest hierarchical elements (joists or beams), moving to higher hierarchical elements (girders) and ultimately to the columns is exact, but tedious, time consuming and errors can accumulate through subsequent steps. Nevertheless, many students continue to use this method. The Müller-Breslau Method is far more efficient and less prone to errors.

Before leaving the method of statics to analyze such grids, consider the slightly more complicated example of inclined structural members known as rafters. These rafters are supported by a vertical wall on one end, and they connect through a hinge at the crown of the roof to a similar set of rafters mirrored about the crown.

Figure 3.16 shows the geometry of one set of rafters supported by a wall. Here, the second mirrored set of rafters is not shown, but rather they are symbolized by hinged, green roller connections

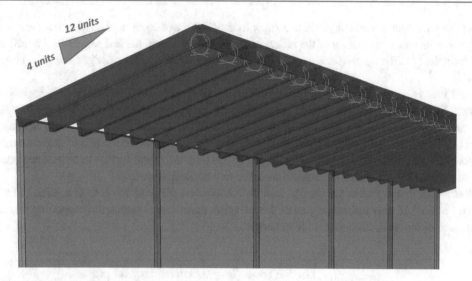

Fig. 3.16 Initial geometry of sloped rafter example

Assume that each rafter is simply supported, i.e. there is a hinged vertical support at each end of the rafter, one support is the vertical wall, the other is the crown ridgeline of the roof. Figure 3.17 shows an alternate view of the problem.

Roof DL=18 force/length^2

Roof LL=30 force/length^2

Wall DL=10 force/length^2

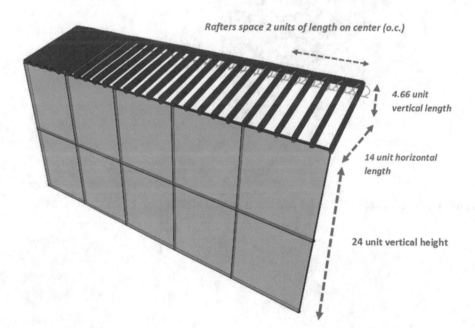

Fig. 3.17 Alternate view of geometry of sloped rafter example

Suppose the rafters were subjected to a snow load of 30 force/length^2 and dead load on the inclined surface that includes self weight of the rafters at 18 force/length^2 and the self weight of the wall which is 10 force/length^2. The goal of this exercise is to find the reaction as a force/length along the bottom of the wall.

Dead load is ***always*** along the length of the member. This is true if the member is horizontal, or inclined. Self weight is always per actual length of the member which makes perfect sense. Yet, it is sometimes convenient to re-calculate dead load as load along a horizontal plane. Why? Because live load (LL) is presented always as if the load were projected on a horizontal plane. Students must be comfortable with both transformations; changing a load on a horizontal surface to an inclined member, as well as changing a load on an inclined member to a horizontal surface.

Figure 3.18 demonstrates the idea of load on a horizontal plane which is then applied to inclined rafters. Figure 3.18 also shows the goal of this exercise, namely to calculate the supporting reaction at the bottom of the wall, expressed as force/length.

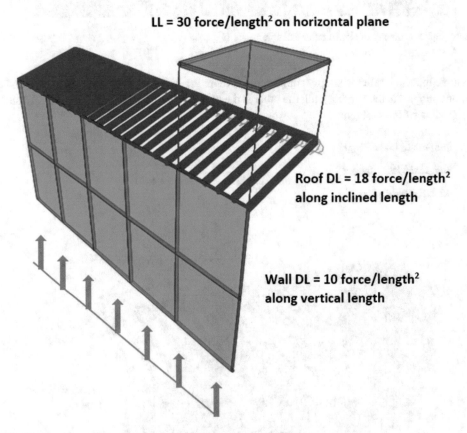

LL = 30 force/length^2 on horizontal plane

Roof DL = 18 force/length^2 along inclined length

Wall DL = 10 force/length^2 along vertical length

Fig. 3.18 Live load is on horizontal plane, dead load is on inclined plane

In this example three loads are studied:

- a snow load of 30 force/length^2
- a dead load on the inclined roof that includes self weight of the rafters at 18 force/length^2
- a dead load of the supporting vertical wall at 10 force/length^2

The goal of this exercise is to calculate the supporting force at the bottom of the wall, this is expressed as force/length. The thickness of the wall is not needed in this problem since its self weight has been given in terms of force/length^2, not force/length^3.

Figure 3.19 shows the transformation of changing a LL which is projected along the horizontal, to a statically equivalent load projected along the rafter incline. The calculations are in Fig. 3.19.

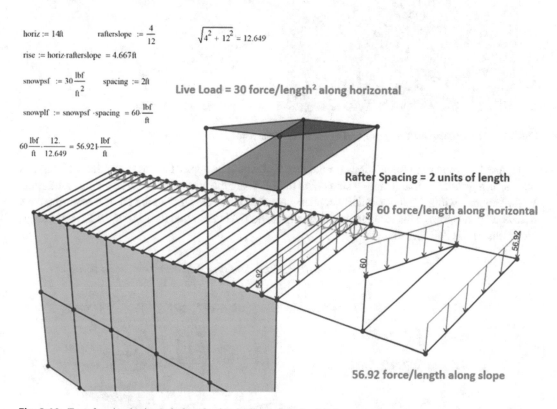

Fig. 3.19 Transforming horizontal plane load to inclined plane load

Conversely, Fig. 3.20 shows the transformation of changing a DL which is always projected along the length of the member, (here the inclined rafter) to a statically equivalent load projected along the horizontal. These calculations are in Fig. 3.20.

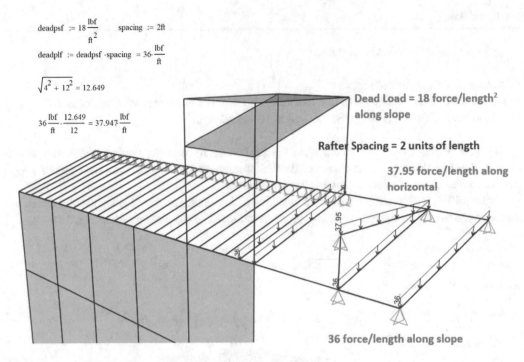

Fig. 3.20 Transforming inclined plane load to horizontal plane load

Figure 3.21 shows the calculations for the load that would be applied to the top of the wall, but this load is "big picture" in that it simply starts with the LL along the horizontal as force/length^2 multiplied by the tributary width. Then it similarly applies the technique to the DL, but the DL is transformed to a horizontal plane prior to the tributary width multiplication. It is clear that the sum of the LL and the DL along this line is 343 force/length.

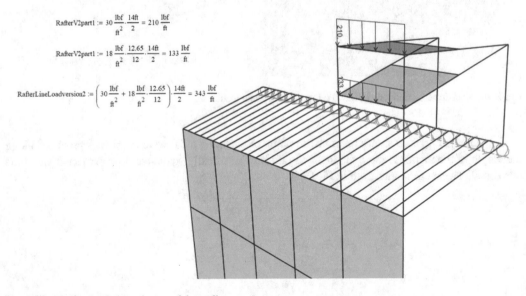

Fig. 3.21 Finding load along the top of the wall

To check the previous value of the line load (force/length) at the top of the wall found to be 343 force/length, the calculations for the reactions at the ends of rafters can be used. Each rafter feels 686 units of force at each end. These loads are divided by the spacing of the supports at the top of the wall, here 2 units of length. Thus the line load (force/length) at the top of the wall can be expressed as was done in Fig. 3.22:

$$56.921 \frac{lbf}{ft} + 36 \frac{lbf}{ft} = 92.921 \cdot \frac{lbf}{ft}$$

$$60 \frac{lbf}{ft} + 37.95 \frac{lbf}{ft} = 97.95 \cdot \frac{lbf}{ft} \qquad Reactionversion1 := \frac{97.95 \frac{lbf}{ft} \cdot 14ft}{2} = 686 \cdot lbf$$

$$RafterLineLoadversion1 := \frac{686 lbf}{2ft} \qquad RafterLineLoadversion1 = 343 \frac{lbf}{ft}$$

Fig. 3.22 Alternate means of finding load along the top of the wall

The load at the top plus the load of the wall itself must be supported by the line load (force/length) at the bottom of the wall. The wall DL was given as 10 force/length^2, and the wall itself is 24 units of length tall. Thus the calculation for the supporting line load reaction at the base is in Fig. 3.23:

$$BottomForceperLength := 343 \frac{lbf}{ft} + 24ft \cdot 10 \frac{lbf}{ft^2} = 583 \cdot \frac{lbf}{ft}$$

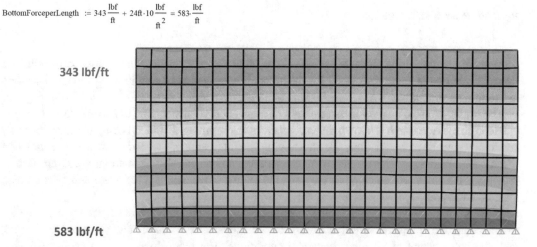

343 lbf/ft

583 lbf/ft

Fig. 3.23 Load flow in vertical wall including self-weight

The Müller-Breslau Method

All statics calculations can be avoided if the Müller-Breslau Method is used. This method is rapid, elegant and exact for all statically determinate structures. It is quick and exact for a grid of beams that are hinged at each end, as the grids are in Timber Design and in Steel Design. Furthermore, the method is exact for gravity loads and for horizontal loads. Finally, the method is extremely fast and precise for statically indeterminate structures, but there it is not exact. For indeterminate analyses, all that will be needed are circular arc lengths and a parametric drawing tool.

The Müller-Breslau Method is so named because it extends the innovative work of one of the world's most brilliant structural analysts. It is curious that Müller-Breslau never finished his university degree! See Fig. 3.24.

Fig. 3.24 Heinrich Müller-Breslau

Many civil engineering students have used the Müller-Breslau method in the context of influence lines. This indeed is the starting point for the broader investigation of equilibrium of determinate and indeterminate structures.

The Müller-Breslau method is best thought of as an energy balance. The unknown reaction sought is forced to move through a known displacement. In the resulting distorted beam, all other boundary conditions must be enforced. Only the effect being sought (reaction, shear or moment) has a violation of boundary conditions. This displacement can be a unit amount, as all deformations are assumed to be "small". If the unknown reaction is a moment, the known movement that the reaction moves through must be a rotation.

When the unknown reaction is displaced some amount, the entire remainder of the structure, be it a simple beam or a complicated grid of beams, deforms in accordance to the known forced movement, with the enforcement of all other boundary conditions. The work done by the unknown moving through the known deformation must balance the work done by all the applied loads moving through their prescribed, associated displacements. Thus Eq. (3.1) summarizes the Müller-Breslau Method:

$$uknown \cdot \Delta = \sum Force_i \cdot \delta_i \qquad (3.1)$$

An elementary statically determinate beam analysis will showcase the method summarized in Eq. (3.1). Then more complicated grids of beams will be studied. Figure 3.25 shows a simply supported beam, each gridline is spaced 5ft. apart. The load is varied but uniform over various portions of the beam.

Fig. 3.25 Simply supported beam with multiple distributed loads

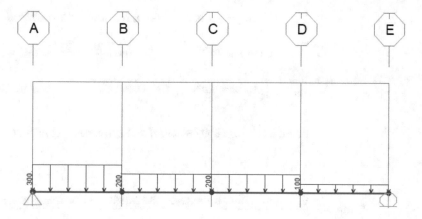

To calculate the right reaction, displace that reaction 1 ft. but maintain all remaining boundary conditions. Here, the remaining boundary conditions are zero vertical movement and zero horizontal movement at Gridline A. A more subtle requirement is that Gridline E does not change position horizontally. Then, measure the displacement at the points of external load application, here the distributed loads are broken up and are concentrated at the centroid of each load distribution. Similar triangles immediately establish the "loft" of each sub-load. These steps are shown in Fig. 3.26.

Fig. 3.26 Lofting right reaction 1 ft. to find support force there

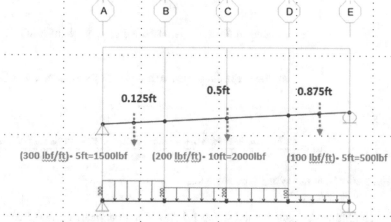

The calculations to find the right reaction are rapid are are evident in Fig. 3.27.

$$\text{LengthAB} := 5\text{ft} \qquad \text{LengthBD} := 10\text{ft} \qquad \text{LengthDE} := 5\text{ft}$$

$$\text{wAB} := 300\frac{\text{lbf}}{\text{ft}} \qquad \text{wBD} := 200\frac{\text{lbf}}{\text{ft}} \qquad \text{wDE} := 100\frac{\text{lbf}}{\text{ft}}$$

$$\text{loftAB} := 0.125\text{ft} \qquad \text{loftBD} := 0.5\text{ft} \qquad \text{loftDE} := 0.875\text{ft}$$

$$\text{RRight} \cdot \Delta = \text{wAB} \cdot \text{LengthAB} \cdot \text{loftAB} + \text{wBD} \cdot \text{LengthBD} \cdot \text{loftBD} + \text{wDE} \cdot \text{LengthDE} \cdot \text{loftDE}$$

$$\Delta := 1\text{ft}$$

$$\text{RRight} := \frac{\text{wAB} \cdot \text{LengthAB} \cdot \text{loftAB} + \text{wBD} \cdot \text{LengthBD} \cdot \text{loftBD} + \text{wDE} \cdot \text{LengthDE} \cdot \text{loftDE}}{\Delta} = 1625\text{lbf}$$

Fig. 3.27 Calculation to find right reaction force

Similarly, the steps needed to find the left reaction are shown in Fig. 3.28.

$$\text{LengthAB} := 5\text{ft} \qquad \text{LengthBD} := 10\text{ft} \qquad \text{LengthDE} := 5\text{ft}$$

$$\text{wAB} := 300\frac{\text{lbf}}{\text{ft}} \qquad \text{wBD} := 200\frac{\text{lbf}}{\text{ft}} \qquad \text{wDE} := 100\frac{\text{lbf}}{\text{ft}}$$

$$\text{loftAB} := 0.875\text{ft} \qquad \text{loftBD} := 0.5\text{ft} \qquad \text{loftDE} := 0.125\text{ft}$$

$$\text{RLeft} \cdot \Delta = \text{wAB} \cdot \text{LengthAB} \cdot \text{loftAB} + \text{wBD} \cdot \text{LengthBD} \cdot \text{loftBD} + \text{wDE} \cdot \text{LengthDE} \cdot \text{loftDE}$$

$$\Delta := 1\text{ft}$$

$$\text{RLeft} := \frac{\text{wAB} \cdot \text{LengthAB} \cdot \text{loftAB} + \text{wBD} \cdot \text{LengthBD} \cdot \text{loftBD} + \text{wDE} \cdot \text{LengthDE} \cdot \text{loftDE}}{\Delta} = 2375\text{lbf}$$

Fig. 3.28 Calculation to find left reaction force

The Müller-Breslau method shines brightest for complicated grids of beams, because the technique is incredibly simple, even for the most complex configurations. All of the calculations could be done on a table napkin, since straight lines always ensue, due to the displacement of a single reaction in a determinate configuration of beams, with all other boundary conditions remaining in place.

Figure 3.29 shows a grid of elements that change directions. The entire grid is subjected to 115 force/length^2. The gaps at the ends of elements in a plan view drawing symbolize the connection as hinged, i.e. no moment is transferred from element to element through such a hinge. The reaction in column B1 is sought.

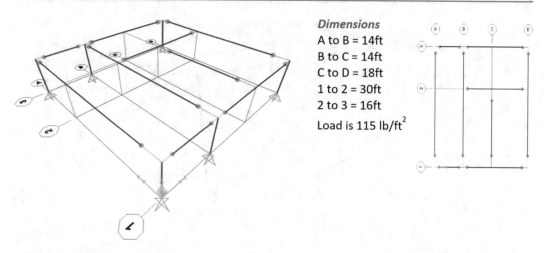

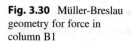

Dimensions
A to B = 14ft
B to C = 14ft
C to D = 18ft
1 to 2 = 30ft
2 to 3 = 16ft

Load is 115 lb/ft^2

Fig. 3.29 Geometry for grid of elements that changes directions

Since the force in Column B1 is sought, the Müller-Breslau method calls for a unit displacement of the reaction at B1 and a subsequent lofting of the entire remainder of the structure. All other boundary conditions must be enforced. This is shown in Fig. 3.30, but no such powerful graphics are needed for the method. A quick sketch is really sufficient.

Fig. 3.30 Müller-Breslau geometry for force in column B1

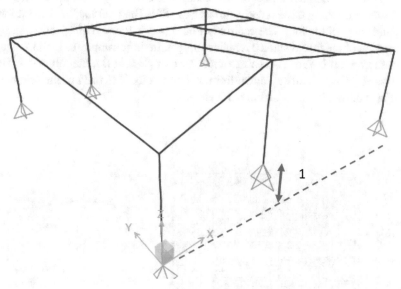

Since the structural grid is statically determinate, all of the lofted lines are straight. Similar triangles are used to immediately identify the loft of the position of each sub-load's centroid. Here, the fold in the form is along Gridline B. Thus drop along Gridline B begins at 1, and ends at 0. Halfway along this length, it drops 0.5. These drops from 1 are clearly designated in Fig. 3.31.

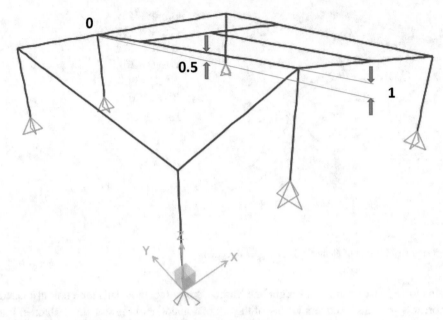

Fig. 3.31 Müller-Breslau initial calculations for force in column B1

Since the original surface folded neatly into two sub-pieces, the centroid of each of the distributed loads will follow these orderly surfaces as well. The total load is divided into two parts, one part on either side of the fold. The centroid of each part is found by inspection, or by similar triangles. Each portion of the folded surface remains straight. In this example, as in most examples studied herein, the resulting loft is ½ of ½ because of similar triangles. At B1, the loft was 1. Halfway down B1 the loft was ½. Moving halfway down from that mid-point is ½ of ½ of the original loft. This is depicted in Fig. 3.32 in plan view and in a 3D view.

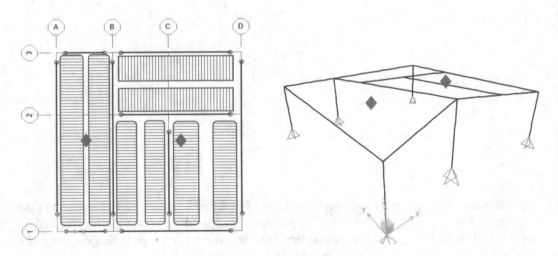

Fig. 3.32 Locating centroids of loads on either side of the fold

Here it is apparent that the loft at the centroid of each sub load is ½ of ½. Notice that the complexity of the inner grid never is needed or calculated with this method. This is shown in Fig. 3.33.

Fig. 3.33 Calculating the loft corresponding to each centroid of load

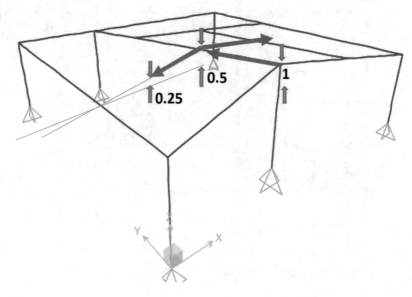

Finding the force in Column B1 is immediate and simple. The steps are described in Fig. 3.34.

$\text{LengthAB} := 14\text{ft}$

$\text{LengthBC} := 14\text{ft}$

$\text{Area1} := (\text{Length12} + \text{Length23}) \cdot \text{LengthAB} = 644\text{ft}^2$

$\text{LengthCD} := 18\text{ft}$

$\text{Area2} := (\text{Length12} + \text{Length23}) \cdot (\text{LengthBC} + \text{LengthCD}) = 1472\text{ft}^2$

$\text{Length12} := 30\text{ft}$

$\text{Length23} := 16\text{ft}$

$\text{yCentroid1} := 0.5 \cdot 0.5 = 0.25 \qquad\qquad \text{yCentroid2} := 0.5 \cdot 0.5 = 0.25$

$\text{DL} := 115\dfrac{\text{lbf}}{\text{ft}^2}$

$\text{FColB1} := \text{DL} \cdot \text{Area1} \cdot \text{yCentroid1} + \text{DL} \cdot \text{Area2} \cdot \text{yCentroid2} = 60835\text{lbf}$

Fig. 3.34 Müller-Breslau final calculations for force in column B1

Extremely complicated floor shapes, with cutouts, openings and variation of loads all present absolutely no additional complexity to the Müller-Breslau method, whereas with a traditional statics approach, this would cause many difficulties.

Figure 3.35 shows one extremely complicated situation. Solving for the column reaction is immediate and simple.

Two Area Loads

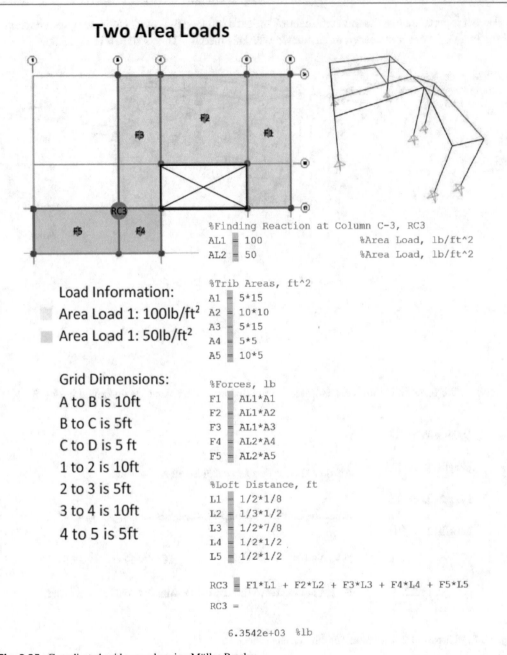

%Finding Reaction at Column C-3, RC3
AL1 = 100   %Area Load, lb/ft^2
AL2 = 50   %Area Load, lb/ft^2

%Trib Areas, ft^2
A1 = 5*15
A2 = 10*10
A3 = 5*15
A4 = 5*5
A5 = 10*5

%Forces, lb
F1 = AL1*A1
F2 = AL1*A2
F3 = AL1*A3
F4 = AL2*A4
F5 = AL2*A5

%Loft Distance, ft
L1 = 1/2*1/8
L2 = 1/3*1/2
L3 = 1/2*7/8
L4 = 1/2*1/2
L5 = 1/2*1/2

RC3 = F1*L1 + F2*L2 + F3*L3 + F4*L4 + F5*L5

RC3 =

6.3542e+03 %lb

Load Information:

Area Load 1: 100lb/ft^2

Area Load 1: 50lb/ft^2

Grid Dimensions:

A to B is 10ft

B to C is 5ft

C to D is 5 ft

1 to 2 is 10ft

2 to 3 is 5ft

3 to 4 is 10ft

4 to 5 is 5ft

Fig. 3.35 Complicated grid example using Müller-Breslau

Figure 3.36 shows another complicated horizontal grid of beams. The Müller-Breslau method easily solves for a column force.

Fig. 3.36 Another grid example using Müller-Breslau

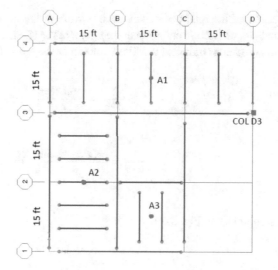

% FORCE WHEN COLUMN D3 LOFTED 1FT:

W1= 100*(lb/ft^2); % A1 & A2 AREA LOAD
W3= 150*(lb/ft^2); % A3 AREA LOAD

FA1= 0.5*0.5*(15*ft*45*ft)*100*(lb/ft^2);
FA2= (1/6)*0.5*(30*ft*15*ft)*100*(lb/ft^2);
FA3= 0.25*0.5*(15*ft*15*ft)*150*(lb/ft^2);

FCol=FA1+FA2+FA3

FCol = (99375*lb)/4
     = 24.84 k

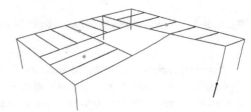

Müller-Breslau Method and Shear at a Point

The technique for calculating equilibrating shear at some point in a beam is exactly the same as the Influence Line method for shear. At the point of investigation of shear, remove the structure's ability to carry shear, and displace it a unit amount. Enforce all other boundary conditions in the distorted shape. Recall the sign convention for internal shear is "down on the left free body, up on the right free body". The sign convention is shown in Fig. 3.37.

Fig. 3.37 Sign convention for positive shear

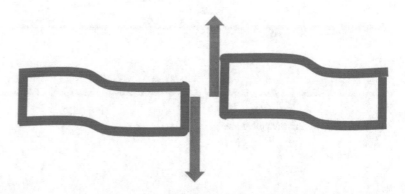

If the cut is at the center of the member's span, each portion of the beam lofts the same amount, namely ½ to create a total gap of 1. If the cut is not at the center of the span, a simple rule is used to find the geometry of the cut. By similar triangles the lifted amount at the cut is described in Fig. 3.38.

Fig. 3.38 Similar triangles to be used in shear calculation

$$\frac{1}{a+b} = \frac{height\ top}{b} \quad \text{thus} \quad \frac{b}{a+b} = height\ top$$

$$\frac{1}{a+b} = \frac{height\ bottom}{a} \quad \text{thus} \quad \frac{a}{a+b} = height\ bottom$$

This rule is shown graphically in Fig. 3.39.

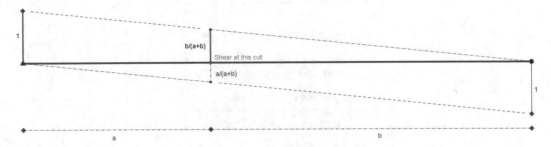

Fig. 3.39 Graphical depiction of similar triangles in shear calculation

This geometric construction is used to find the loft of the distorted beam at the point of load application. Notice the opposing signs, some lofts are positive and some are negative. To find the shear at a cut at a known location due to loads at other known locations, use the rule shown in Fig. 3.40.

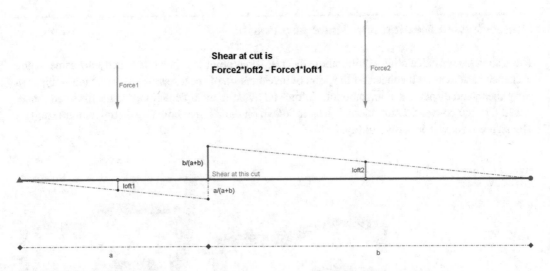

Fig. 3.40 Finding loft of applied loads

Of course, if there are more loads applied, simply add them to the summation.

More complicated grid layouts, with variations of load, do not add any complexity to the Müller-Breslau method. Figure 3.41 shows a complex layout. This example will demonstrate the ease with which the calculations are performed.

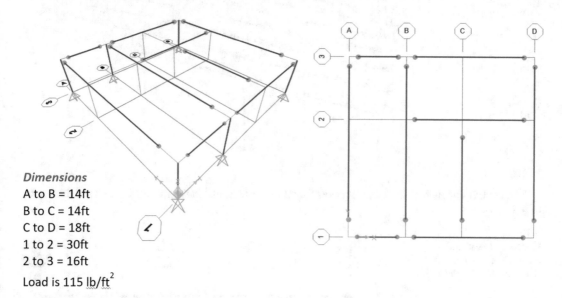

Dimensions

A to B = 14ft

B to C = 14ft

C to D = 18ft

1 to 2 = 30ft

2 to 3 = 16ft

Load is 115 lb/ft^2

Fig. 3.41 Geometry of complicated grid example

To find the shear transferred from the beam on Gridline 2 to the beam on Gridline B, remove the capabilities of that connection to carry shear. Then displace the connection in the shear direction a net differential of 1 ft and enforce all other boundary conditions. Here the "top" is 1 and the "bottom" is zero as described in Fig. 35 because the cut is so close to the beam on Gridline B that the distance "a" is zero.

Notice the original area East of Beam B folds along the Beam 2 "ridge line". This gives you a hint that you need to find the centroid of each part of the folded area. Then another fold occurs along the Beam C "ridge line". Thus, three centroids are shown with a red diamond in Fig. 3.43, one for of each sub-area.

Key Idea *When a surface is folded, the sub-areas of load must be individually found on either side of the fold. Do not ever take a distributed load over a fold and find its centroid, always divide the area into pieces on either side of the fold.*

Also notice that the area left of Beam B does not fold, thus it cannot affect the shear force passing through this connection. All of these issues are evident in Fig. 3.42

Fig. 3.42 Müller-Breslau
geometry to find shear at
connection 2B

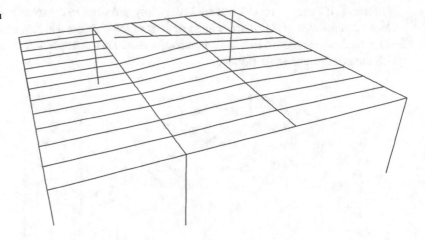

The centroid of each of these sub-areas is found by inspection. These are shown with red diamonds in Fig. 3.43.

Fig. 3.43 Locating
centroids of loads on either
side of the fold for shear 2B

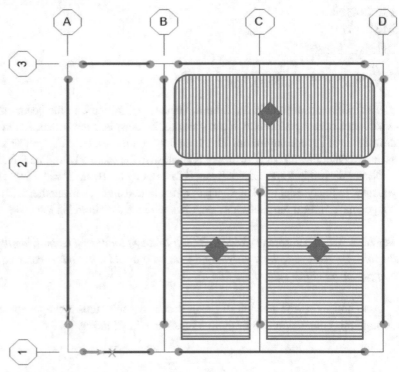

These centroids are located by inspection as the centers of the rectangles. The loft associated with each centroid is found by similar triangles. In Fig. 3.44, the solid green arrows denote the drops from the starting point of 1. The point YStep1 is found by similar triangles. The subsequent ½ and ½ drops are truly driven by similar triangles, yet the calculation is trivial because the path from the point YStep1 to the point YStep2 drops half its height. Similarly, the path from point YStep2 to the point Area2Lofted is yet another 50% drop.

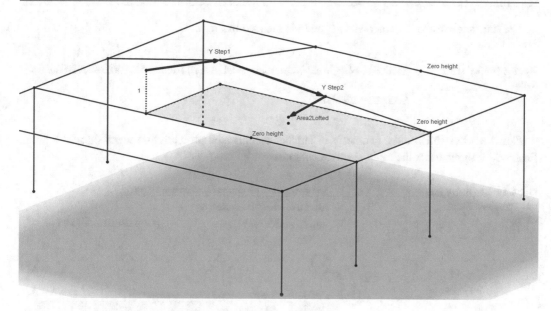

Fig. 3.44 Similar triangles depicted on grid

It may not be necessary to describe similar triangles, but students find the following trick easy to remember, rise1/run1 = rise2/run2. This is demonstrated in Fig. 3.45.

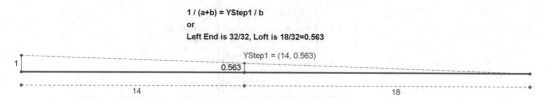

Fig. 3.45 Useful tool for similar triangle

Note that in the previous example the starting point is 1 because the shear cut is so very close to the connecting beam that all the lift goes into the "top", not the "bottom" of the broken curve shown previously in Fig. 3.40.

The calculations to find the shear are rapid. The are depicted in Fig. 3.46.

$\text{LengthAB} := 14\text{ft}$

$\text{LengthBC} := 14\text{ft}$

$\text{LengthCD} := 18\text{ft}$ $\text{yCentroid1} := 0.5 \cdot 0.5 = 0.25$

$\text{Length12} := 30\text{ft}$
$\quad \text{Area1} := (\text{LengthBC} + \text{LengthCD}) \cdot \text{Length23} = 512\,\text{ft}^2 \qquad \text{yStep1} := \dfrac{1}{\text{LengthBC} + \text{LengthCD}} \cdot \text{LengthCD} = 0.5625$

$\text{Length23} := 16\text{ft}$

$\text{DL} := 115\dfrac{\text{lbf}}{\text{ft}^2} \qquad \text{Area2} := \text{LengthBC} \cdot \text{Length12} = 420\,\text{ft}^2 \qquad \text{yStep2} := \dfrac{1}{2} \cdot \text{yStep1} = 0.2813$

$\qquad\qquad \text{Area3} := \text{LengthCD} \cdot \text{Length12} = 540\,\text{ft}^2 \qquad \text{yCentroid2} := \dfrac{1}{2} \cdot \text{yStep2} = 0.1406 \qquad\qquad \text{yCentroid3} := \dfrac{1}{2} \cdot \text{yStep2} = 0.1406$

Fig. 3.46 Calculations for shear example

The final shear force is immediately found via the step in Fig. 3.47.

Fig. 3.47 Final shear force

$$\text{Shear2B} := DL \cdot Area1 \cdot yCentroid1 + DL \cdot Area2 \cdot yCentroid2 + DL \cdot Area3 \cdot yCentroid3$$

$$\text{Shear2B} = 30245 \text{lbf}$$

Figure 3.48 establishes the geometry of another complicated problem that seeks shear at a point. Figure 3.49 summarizes the rapid shear calculation.

Fig. 3.48 Geometry for more complex shear example

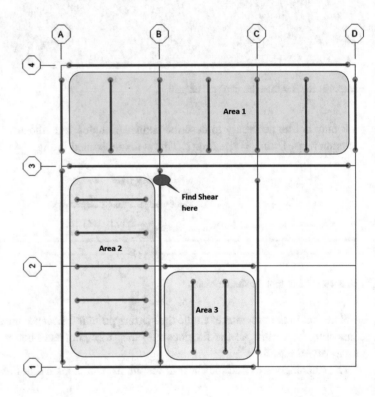

Grid spacing A thru D = 15ft, Grid spacing 1 thru 4 = 15ft

Area 1 Load = Area 2 Load = 100 lb/ft²

Area 3 Load = 150 lb/ft²   Find Shear just south of B-3

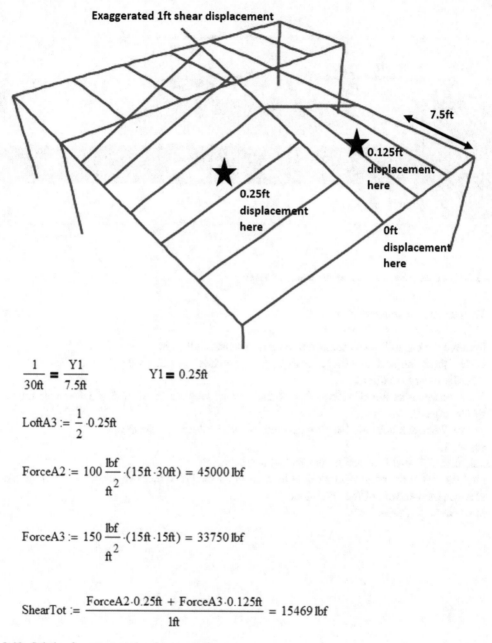

$$\frac{1}{30\text{ft}} = \frac{Y1}{7.5\text{ft}} \qquad Y1 = 0.25\text{ft}$$

$$\text{LoftA3} := \frac{1}{2} \cdot 0.25\text{ft}$$

$$\text{ForceA2} := 100\,\frac{\text{lbf}}{\text{ft}^2} \cdot (15\text{ft} \cdot 30\text{ft}) = 45000\,\text{lbf}$$

$$\text{ForceA3} := 150\,\frac{\text{lbf}}{\text{ft}^2} \cdot (15\text{ft} \cdot 15\text{ft}) = 33750\,\text{lbf}$$

$$\text{ShearTot} := \frac{\text{ForceA2} \cdot 0.25\text{ft} + \text{ForceA3} \cdot 0.125\text{ft}}{1\text{ft}} = 15469\,\text{lbf}$$

Fig. 3.49 Solution for more complex shear example

To find the bending moment at some cut in the beam, remove the capabilities of the beam to carry moment at that cut. Then, for the cut beam, ensure that the relative rotation between the beam on either side of the cut is $\theta = 1$. This is startling the first time it is explored, since $\theta = 1$ is not 1 degree, nor 1 radian, simply 1. Furthermore, we show it "zoomed in", i.e. it appears large, but in fact, it is small. As before, a key idea is to break up distributed loads into sub-areas on either side of the fold. $\theta = 1$ is simpler to understand if one portion of the beam remains horizontal, for example if the moment is sought for a cut on a cantilever beam, one side of the cut remains horizontal. This is shown in Fig. 3.50 where a uniform load is applied over a portion of the beam, and a point load is applied at the free tip. The moment at $x=7$ is sought.

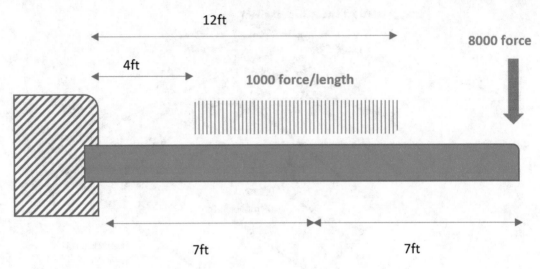

Fig. 3.50 Geometry and loads of cantilever beam example

To calculate the moment at the cut:

- Remove the capabilities of the beam to carry moment at the cut
- Distort the beam such that $\theta=1$, the angle that is formed on either side of the cut, but maintain all other boundary conditions
- Yet the angle is assumed be "small" such that any arc length s, must equal the distance r multiplied by the angle θ.
- $s = r \cdot \theta$ where s is an arc length measured as a vertical line for small θ
- $s = r \cdot 1$
- Establish the height formed by the kinked beam triangle
- Find the loft at the point of centroidal load, or point load on either side of the fold, do not use the smeared out centroid of the entire load
- $M \text{ at cut} = \sum force_i \cdot loft_i$

The geometrical calculations are shown in Fig. 3.51.

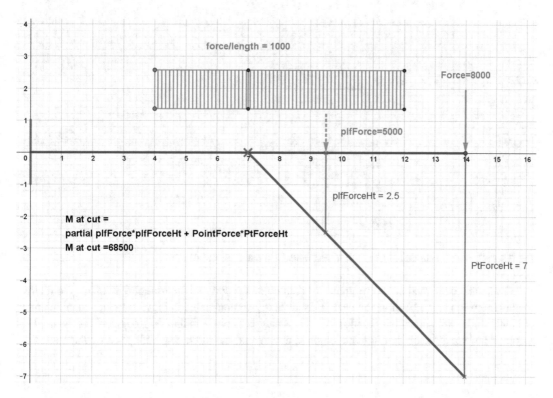

Fig. 3.51 Müller-Breslau geometry to find shear at x=7ft

Notice in Fig. 3.51 that the load is broken up into two sub-portions. The portion to the right of the cut is 5 units of length, at 1000 force/length totaling 5000 units of force. This is applied at the centroid of that portion to the right of the cut, namely at 2.5 units of length to the right of the cut. It would have been completely incorrect to take the entire load of 8000 and to apply that $x=8$ which is the centroid of the entire distributed load. It is correct to sub divide the load into two portions, left and right of the fold. Here the portion left of the fold does no work.

If there are no horizontal portions of the beam on either side of the cut, as in the case of a simply supported beam, then the relative rotation between the segments across the cut is still $\theta=1$, but to see that clearly, first find the height of the kink in the deformed beam. Then, if there is a distributed load find the centroid on either side of the fold.

For example, a simply supported beam is shown in Fig. 3.52. It is 32 units long, and the moment at $x=22$ is sought. Deform the beam as shown in Fig. 3.52 using the height at the kink to be:

$$Height\ at\ cut = \frac{a \cdot b}{a+b} \tag{3.2}$$

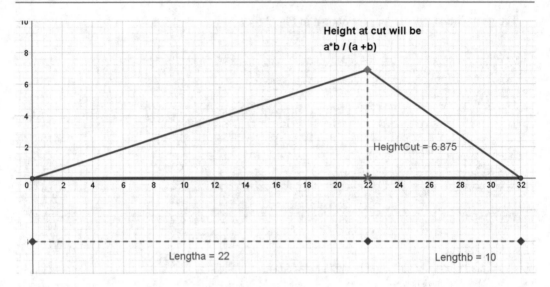

Fig. 3.52 Geometric rule for height of loft to find moment at cut

Why is this deformed shape so precisely defined at the apex of the triangle? Because the relative rotation between segments must be $\theta=1$. This is clearly demonstrated in Fig. 3.53. Such manipulations of small angles are somewhat paradoxical. The value is 1, not 1 radian or 1 degree. Yet the angle is assumed be "small" such that any arc length s, must equal the distance r multiplied by the angle θ.

$$s = r \cdot \theta$$

$$s = r \cdot 1$$

If this is difficult to grasp, that is normal. This same topic will return when rotation of rigid roofs is studied. For now, simply use the construction of Fig. 3.52. The logic of Fig. 3.52 will be explained in the following two figures.

In Fig. 3.53, if $\theta=1$, then the arc length, or analogous vertical length at some distance r is equal r. Thus swinging out the radial distance b, provides the height b. Or in the other direction, swinging some radius a gives the height a.

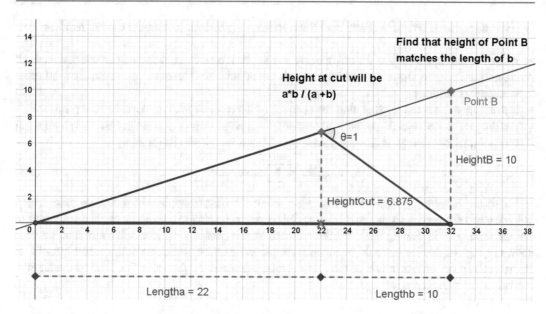

Fig. 3.53 First explanation for height of loft to find moment at cut

Yet another way of looking at this same shape is to view the left side of the construction. This is shown in Fig. 3.54.

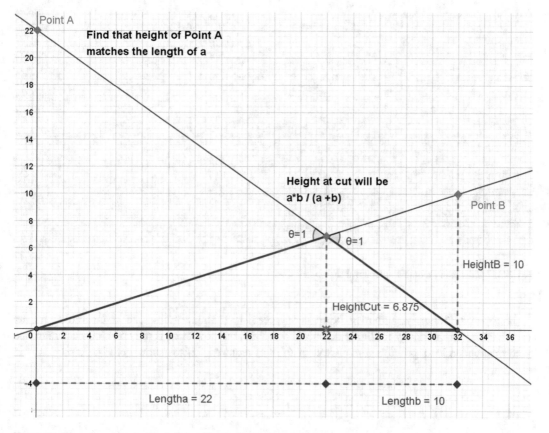

Fig. 3.54 Second explanation for height of loft to find moment at cut

As with reactions and with shear, the Müller-Breslau method for moment at a point requires:

- removing the capabilities of the effect being sought, be it a reaction, a shear or an internal moment
- create the distorted shape by moving through the effect a unit amount yet maintaining all other boundary conditions
- if distributed loads are applied, find the centroid of the original load on the un-lofted form
- multiply the concentrated applied load (either from a point load or as a centroidal load) by the loft that point moves though in the distorted shape. That is the effect sought then:

$$1 \cdot \textit{effect} = \sum_i \textit{force}_i \cdot \textit{loft}_i \qquad (3.3)$$

For example, a simply supported beam subjected to a uniformly applied load over its entire length will have two sub pieces. Compare this to the same beam subjected to a centrally applied point load to see the difference in dealing with distributed loads versus point loads. The uniformly load beam is shown in Fig. 48. Notice it would be incorrect to place the entire $w*L$ force at the center loft of $L/4$. Rather, sub-forces on either side of the fold must be calculated as shown in Fig. 3.55, providing the answer $wL^2/8$ as required.

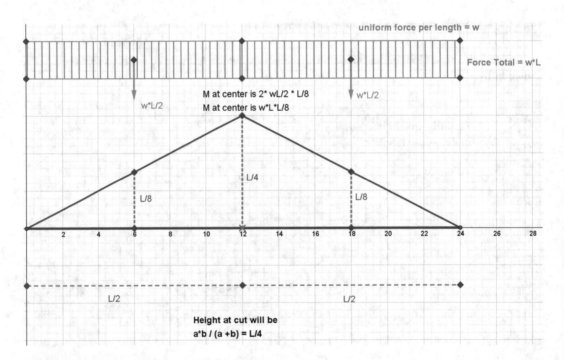

Fig. 3.55 Calculating moment at a cut for distributed load

If however, a simply supported beam is subjected to a single point load at the mid-span, the calculations work perfectly for the folded beam as before, but now there is only one loft needed in line with the applied load, providing an answer of $PL/4$ as required. This is shown in Fig. 3.56.

Fig. 3.56 Calculating moment at a cut for point load

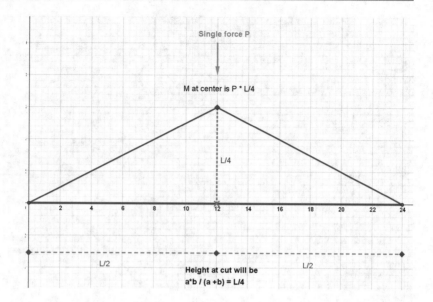

A slightly more complicated numerical example follows. In Fig. 3.57, a beam has the following:

- Gridline spacing A to B and B to C is 8 units of length
- Gridline spacing C to D is 12ft units of length
- Gridline spacing D to E is 8ft units of length
- One concentrated load of 20,000 units of force is applied midway between lines D and E
- One distributed load of 800 force/length is applied uniformly between lines B and D
- There is an internal hinge at line C which transfers shear, but not moment
- Pinned at line A, rollers at lines D and E
- Find the moment in the beam at Gridline D

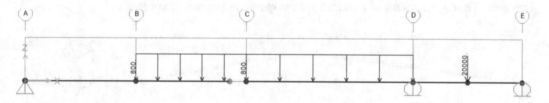

Fig. 3.57 Geometry of determinate beam with internal hinge example

The geometrical construction used to analyze for the moment at D first requires the removal of moment carrying capabilities at D. A positive, relative rotation between portions DC and DE of the cracked beam ensues. Portion DE must remain horizontal since the prompt was to find the moment at Gridline D, the cut is directly above the roller at D and none of the rollers can move vertically. All other boundary conditions must be enforced. Notice that the hinge at Gridline C allows the beam to kink again. This is shown in Fig. 3.58.

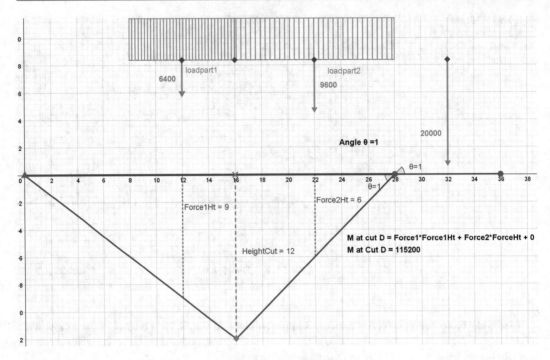

Fig. 3.58 Moment at cut of determinate beam with internal hinge example

Notice that the 20,000 force concentrated at $x=32$ does not induce any moment at D. Is that surprising? Consider the deflected shape of this beam if it were subjected to only the 20,000 force concentrated load applied mid-way between D and C, and no distributed load anywhere else. Would the concentrated force induce bending at D? The presence of the hinge is key in sketching the deformed shape. This is shown in Fig. 3.59. Notice there is no bending, no shear, no reactions left of the interior roller at D. This highlights a fundamental mechanics Truth, bending is directly proportional to curvature. The displacements of the beam are a response to an induced curvature.

Fig. 3.59 Bending is directly proportional to curvature, not to deflection

Two more examples of moment calculation at a particular cut follow. One is for a simply supported beam where the geometric construction $ab/(a+b)$ must be formed. This is shown in Fig. 3.60. In Fig. 3.60, a simply supported beam is 20 units of length. It is subjected to a uniformly applied load of 750 force/length. The moment at a cut $x=8$ is sought.

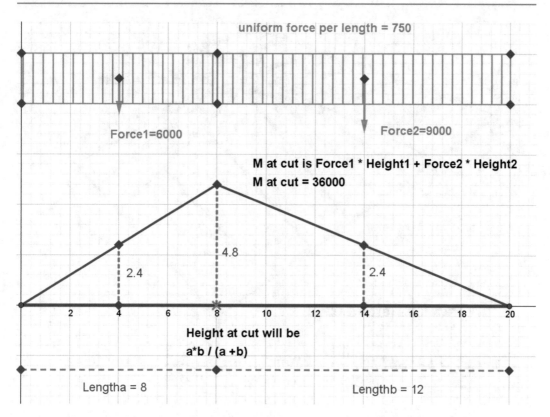

Fig. 3.60 Example, beam with distributed and point loads

A more complicated example which finds moment in one cut for a grid of beams is shown in Fig. 3.61.

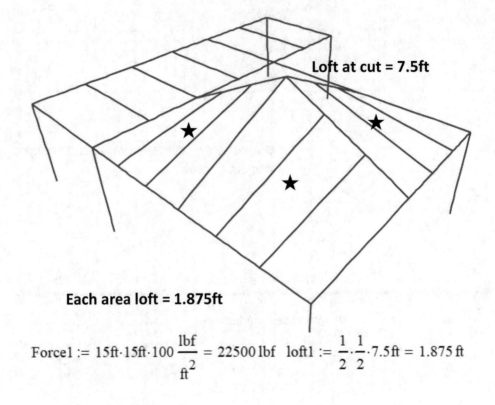

$$\text{Force1} := 15\text{ft} \cdot 15\text{ft} \cdot 100 \, \frac{\text{lbf}}{\text{ft}^2} = 22500 \, \text{lbf} \quad \text{loft1} := \frac{1}{2} \cdot \frac{1}{2} \cdot 7.5\text{ft} = 1.875 \, \text{ft}$$

$$\text{Force2} := 15\text{ft} \cdot 15\text{ft} \cdot 100 \, \frac{\text{lbf}}{\text{ft}^2} = 22500 \, \text{lbf} \quad \text{loft2} := \frac{1}{2} \cdot \frac{1}{2} \cdot 7.5\text{ft} = 1.875 \, \text{ft}$$

$$\text{Force3} := 15\text{ft} \cdot 15\text{ft} \cdot 150 \, \frac{\text{lbf}}{\text{ft}^2} = 33750 \, \text{lbf} \quad \text{loft3} := \frac{1}{2} \cdot \frac{1}{2} \cdot 7.5\text{ft} = 1.875 \, \text{ft}$$

$$\text{MomentAtB2} := \text{Force1} \cdot \text{loft1} + \text{Force2} \cdot \text{loft2} + \text{Force3} \cdot \text{loft3} = 147656 \, \text{ft} \cdot \text{lbf}$$

Fig. 3.61 Moment calculations in complicated grid of beams

Projects

The following projects are meant to be programmed in any environment, such as MathCAD, MATLAB, Grasshopper or GeoGebra. Insights arise from manipulations of the input variables and the study of the effects these changes induce.

Project 3–1 A horizontal grid of beams shown in 2D and in 3D supports only DL= 100lb/ft^2 which includes all self weight. Joists are not shown, but the direction they span is indicated with the arrow. There is an opening in the floor between lines 3 and 4 and C and D.

Find the moment in the beam along Grid1 at point C, i.e. find the moment at C1. The plan view is shown in Fig. 3.62, the 3D view is shown in Fig. 3.63 and the grid spacing is shown in Fig. 3.64.

Fig. 3.62 Project 3–1 geometry, plan view

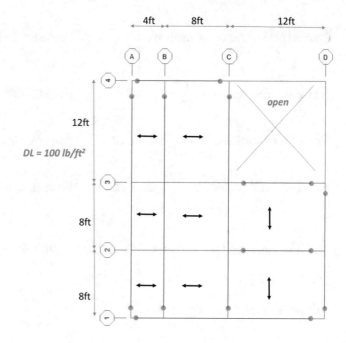

Fig. 3.63 Project 3–1 geometry, 3D view

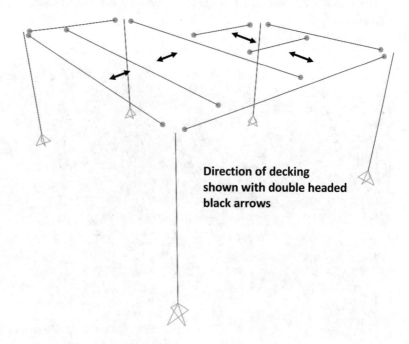

Direction of decking shown with double headed black arrows

$$\text{HeightAtCut} := \frac{12\text{ft} \cdot 12\text{ft}}{12\text{ft} + 12\text{ft}} = 6\,\text{ft}$$

Centroid of Left Area is 14ft north of C1   **Centroid of Left Area is 8ft north of C1**

$$\text{HeightAt14ftNorthofC1} := \frac{6\text{ft}}{28\text{ft}} \cdot 14\text{ft} = 3\,\text{ft} \qquad \text{HeightAt8ftNorthofC1} := \frac{6\text{ft}}{28\text{ft}} \cdot 20\text{ft} = 4.286\,\text{ft}$$

$$\text{HeightAtLeftAreaCentroid} := \frac{1}{2} \cdot 3\text{ft} = 1.5\,\text{ft} \qquad \text{HeightAtRightAreaCentroid} := \frac{1}{2} \cdot 4.286\text{ft} = 2.143\,\text{ft}$$

$$\text{ForceLeft} := (28\text{ft} \cdot 12\text{ft}) \cdot 100\,\frac{\text{lbf}}{\text{ft}^2} = 33600\,\text{lbf} \qquad \text{ForceRight} := (16\text{ft} \cdot 12\text{ft}) \cdot 100\,\frac{\text{lbf}}{\text{ft}^2} = 19200\,\text{lbf}$$

$$\text{MatC1} := \text{ForceLeft} \cdot \text{HeightAtLeftAreaCentroid} + \text{ForceRight} \cdot \text{HeightAtRightAreaCentroid}$$

$$\text{MatC1} = 91546\,\text{lbf} \cdot \text{ft}$$

Fig. 3.64 Project 3–1 geometric calculations

The bending moment at C1 is quickly found via the Müller-Breslau method. This is shown in Fig. 3.65.

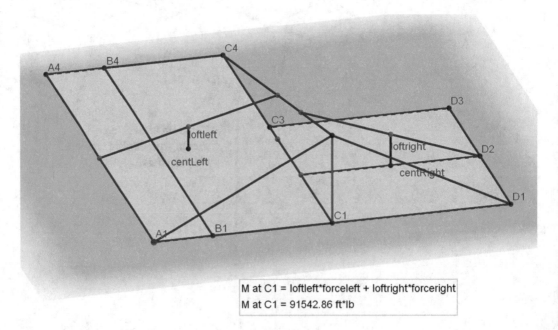

M at C1 = loftleft*forceleft + loftright*forceright
M at C1 = 91542.86 ft*lb

Fig. 3.65 Bending moment found at C1

Project 3–2 Given the following building. Grid spacing between A, B, C, D and E is 34ft and Grid spacing between 1, 2, 3 and 4 is 22ft. The roof feels a uniform Dead Load (DL) of 9lb/ft². Notice all the shorter beams are hinged at their ends, i.e. they do not transfer moment at their ends. Notice the longer beams are continuous through the points where the beams connect, but the longer beams are also hinged at their ends. This is shown in Fig. 3.66.

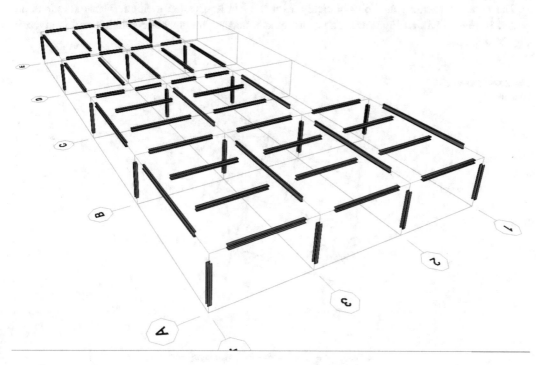

Fig. 3.66 Project 3–2 geometry

The support conditions are extremely clear when viewing the bending moment diagrams shown in Fig. 3.67

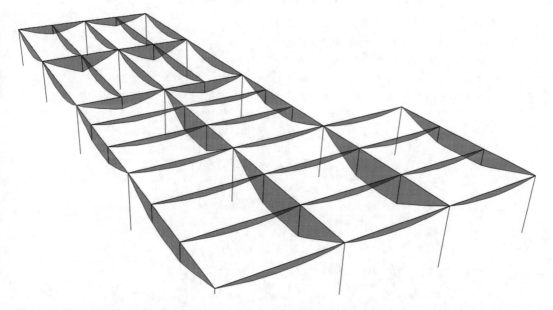

Fig. 3.67 Project 3–2 boundary conditions as seen through bending moment diagrams

Find:

- Column reaction at B3
- Column reacation at B2
- Worst moment on the girder on Grid2 between Grids A and B

The column reactions are found quickly with the Müller-Breslau method. These solutions are posted in Figs. 3.68 and Fig. 3.69. The moment is found via an algebraic solution and is given in Fig. 3.70.

Fig. 3.68 Project 3–2
column B3 force

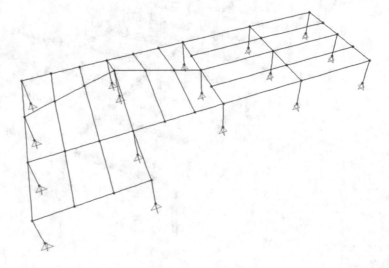

$$B3Force := \frac{1}{2} \cdot \frac{1}{2} \cdot (floorload \cdot Spacing23 \cdot SpacingAB) \ldots$$
$$+ \frac{1}{2} \cdot \frac{1}{2} \cdot (floorload \cdot Spacing34 \cdot SpacingAB) \ldots$$
$$+ \frac{1}{2} \cdot \frac{1}{2} \cdot (floorload \cdot Spacing23 \cdot SpacingBC) \ldots$$
$$+ \frac{1}{2} \cdot \frac{1}{2} \cdot (floorload \cdot Spacing34 \cdot SpacingBC)$$

$$B3Force = 6749.111 \, lbf$$

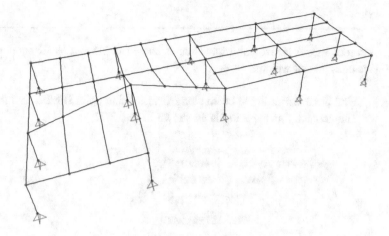

Fig. 3.69 Project 3–2 column B2 force

$$B2Force := \frac{1}{2} \cdot \frac{1}{2} \cdot (floorload \cdot Spacing23 \cdot SpacingAB) \dots$$
$$+ \frac{1}{2} \cdot \frac{1}{2} \cdot (floorload \cdot Spacing23 \cdot SpacingBC) \dots$$
$$+ \frac{1}{2} \cdot \frac{1}{2} \cdot (floorload \cdot Spacing12 \cdot SpacingAB)$$

$$B2Force = 5061.833 \, lbf$$

Fig. 3.70 Project 3–2 M peak on Grid2

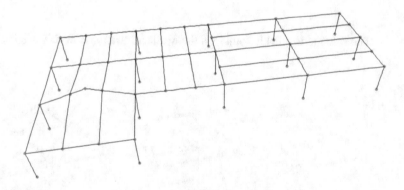

$$arealofted := Spacing12 \cdot \frac{SpacingAB}{2} \qquad loft := 8.5ft$$

$$momentgirder := 4 \cdot \left(\frac{1}{2} \cdot \frac{1}{2} \cdot loft \cdot arealofted \cdot floorload \right)$$

$$momentgirder = 28683.722 \, lbf \cdot ft$$

An algebraic approach with point loads flowing into girder gives:

$$ReactionShort := \frac{plfshort \cdot Spacing12}{2} = 1122 \, lbf$$

$$PointLoad := ReactionShort \cdot 2 = 2244 \, lbf$$

$$Mpeak := PointLoad \cdot \frac{SpacingAB}{3} = 25432 \, lbf \cdot ft$$

Project 3–3 A horizontal grid of beams shown in 2D and in 3D supports only DL= 50lb/ft² which includes all self weight.

Find the moment in the beam along Gridline 3 at point B3, i.e. find the moment at B3.
The geometry and spacing is all shown in Fig. 3.71

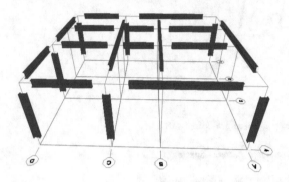

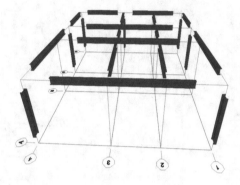

Grid Spacing:

A to B = 14ft	1 to 2= 8ft
B to C = 8ft	2 to 3 = 8ft
C to D = 12ft	3 to 4 = 12ft

If the roof feels 50 lb/ft², find the moment in the beam along Grid Line 3 at point B3

Fig. 3.71 Project 3–3 geometry and spacing

The solution is quickly found via the Müller-Breslau Method. The solution is shown in Fig. 3.72

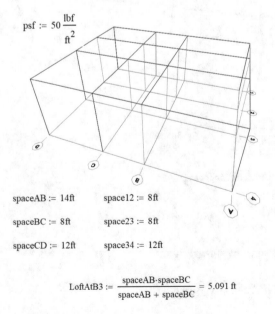

$$\text{psf} := 50\,\frac{\text{lbf}}{\text{ft}^2}$$

spaceAB := 14ft   space12 := 8ft

spaceBC := 8ft   space23 := 8ft

spaceCD := 12ft   space34 := 12ft

$$\text{LoftAtB3} := \frac{\text{spaceAB} \cdot \text{spaceBC}}{\text{spaceAB} + \text{spaceBC}} = 5.091\,\text{ft}$$

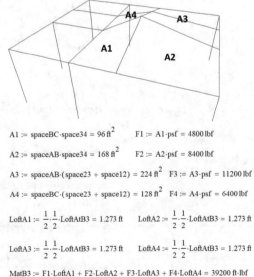

$A1 := \text{spaceBC} \cdot \text{space34} = 96\,\text{ft}^2$ $\quad$ $F1 := A1 \cdot \text{psf} = 4800\,\text{lbf}$

$A2 := \text{spaceAB} \cdot \text{space34} = 168\,\text{ft}^2$ $\quad$ $F2 := A2 \cdot \text{psf} = 8400\,\text{lbf}$

$A3 := \text{spaceAB} \cdot (\text{space23} + \text{space12}) = 224\,\text{ft}^2$ $\quad$ $F3 := A3 \cdot \text{psf} = 11200\,\text{lbf}$

$A4 := \text{spaceBC} \cdot (\text{space23} + \text{space12}) = 128\,\text{ft}^2$ $\quad$ $F4 := A4 \cdot \text{psf} = 6400\,\text{lbf}$

$\text{LoftA1} := \frac{1}{2} \cdot \frac{1}{2} \cdot \text{LoftAtB3} = 1.273\,\text{ft}$ $\quad$ $\text{LoftA2} := \frac{1}{2} \cdot \frac{1}{2} \cdot \text{LoftAtB3} = 1.273\,\text{ft}$

$\text{LoftA3} := \frac{1}{2} \cdot \frac{1}{2} \cdot \text{LoftAtB3} = 1.273\,\text{ft}$ $\quad$ $\text{LoftA4} := \frac{1}{2} \cdot \frac{1}{2} \cdot \text{LoftAtB3} = 1.273\,\text{ft}$

$\text{MatB3} := \text{F1} \cdot \text{LoftA1} + \text{F2} \cdot \text{LoftA2} + \text{F3} \cdot \text{LoftA3} + \text{F4} \cdot \text{LoftA4} = 39200\,\text{ft} \cdot \text{lbf}$

Fig. 3.72 Project 3–3 Moment at B3 solution

Dead Load and Live Load

4

Iterative Design Ideas

When drawing structural framing plans there is a hierarchy of line weight which helps communicate information clearly and rapidly.

1. Columns – Heavy
2. Truss – Heavy Dashed Line
3. Secondary Framing – Typically these are beams with a group designation (B1, B2 etc)
4. Diagonal Brace – Medium Dashed Line
5. Gridlines – Light
6. Dimensions – Light
7. Title and Scale

One can certainly begin with estimates of floor loads, but if one is designing beams, the weight of the beams has not yet been incorporated into the design. Self weight is Dead Load and while it may be small for timber, it is large for reinforced concrete, and it still must be accounted for.

Suppose the design of a simple rectangular building was sought, such as the one shown in Fig. 4.1. How to account for the self weight?

© Springer Nature Switzerland AG 2020
E. Saliklis, *Structures: A Studio Approach*, https://doi.org/10.1007/978-3-030-33153-5_4

Fig. 4.1 Design problem,
elementary structure

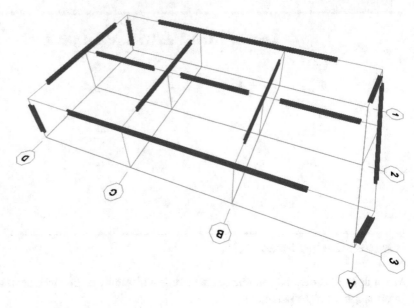

Suppose the elements along gridlines 1 and 3 support the elements along gridlines B and C. The elements along 1 and 3 have a higher hierarchy, namely they support other elements, as well as themselves.

Next, assume that the floor load, the Mechanical Electrical Plumbing (MEP) and other miscellaneous loads all add up to 60 lb/ft^2. Assume the elements are timber, but keep the depth of all the elements controlled due to architectural constraints. Assume these are rectangular cross sections and check the stress of the elements using the simple formula:

$$\sigma = \frac{M}{S} = \frac{M}{b \cdot d^2/6} \tag{4.1}$$

If the stress exceeds some allowable stress, one must redesign, making the section bigger, but then the self weight of course increases. Thus, the design iterates till it satisfies stress and architectural constraints. To create such a simple structure in SAP200 as shown in Fig. 4.2, follow these steps: Define > Section Properties > Frame Sections > Add New Property > Concrete. Even for timber, this is the easiest way to introduce a rectangular cross section. Assign > Frame Releases as appropriate to your model, turn off both strong and weak moments in 3D and only the 3-3 moment in a 2D problem.

Fig. 4.2 Elementary structure in SAP2000 with no moment releases

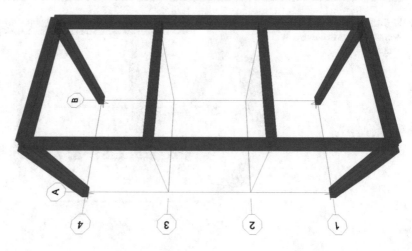

In this timber design example, it is acceptable to assume that structural elements are all simply supported, i.e. the elements are statically determinate. This makes for the easy and quick calculation of peak moment due a uniformly applied load w:

$$M_{\max\ uniform\ load} = \frac{w \cdot L^2}{8} \tag{4.2}$$

And in this example for timber members:

$$\sigma_{allowable} = 900^{lb}/_{in^2} = 6.2 MPa \tag{4.3}$$

Tasks

- Calculate the final force/length^2 which includes self weight and any additional floor loads. But use ft as the length, not in.
- Take the force/length^2 and make it a force/length on the smaller elements
- Check the stress on the smaller elements
- Notice that the smaller elements hit the larger elements as point loads
- Check the stress on the larger elements
- Calculate the final load on the columns coming from the larger elements on Gridlines 1 and 2 as well as from the shorter elements on Gridlines A and D. Notice, two elements attach to each column. This is highlighted in Fig. 4.3.
- COMPARE THIS COLUMN LOAD to the load that would arise from ¼ of the entire roof load, namely one quarter of the TOTAL WEIGHT OF ALL THE BEAMS, the TOTAL WEIGHT OF THE FLOOR

Fig. 4.3 Long and short
beams connect to column at
corners

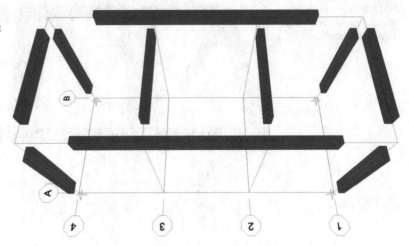

Hint The self weight of timber beams can be expressed first as 35 lbf/ft^3, then multiply by the width and by the height to get lbf/ft or force/length. This is useful to take an initial estimate of dead load.

Project 4–1 Task Design a one story timber structure. There must be one bay along one axis and four or five or six smaller elements along the other axis. In addition to beam self weight, include some additional floor load that is reasonable. Do the steps listed above, and the final check is important, it solidifies your understanding of the gravity load flow.

Figure 4.4 shows an example structure that could have been designed for such a project.

Fig. 4.4 Typical Project
4–1 layout

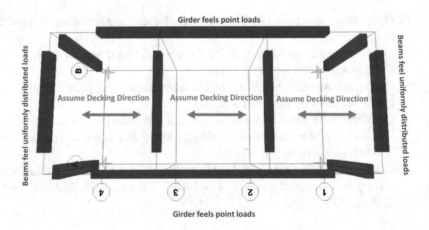

The shorter beams in Fig. 4.4 are 10ft long and the longer beams are 24ft long. Assume the additional floor load is 25lb/ft^2. Notice how the symbol for a hinged connection appears as a gap between elements, which is a standard drawing icon for beams that are statically determinate. The architect requires that the shorter beams do not exceed 10inches of depth. The allowable stress is as before, namely 900lb/in^2 or 6.2MPa.

Initial calculations are shown in Fig. 4.5.

Fig. 4.5 Initial
Project 4–1 calculations

$$\sigma\text{allowable} := 900 \frac{\text{lbf}}{\text{in}^2} = 6.205 \times 10^6 \, \text{Pa}$$

$$L12 := 10\text{ft} \qquad LAD := 24\text{ft}$$

$$\text{FloorLoad} := 25 \frac{\text{lbf}}{\text{ft}^2} \qquad DL2by8 := 2.6 \frac{\text{lbf}}{\text{ft}}$$

$$\text{BeamTribWidth} := \frac{\frac{LAD}{3}}{2} \cdot 2 = 8 \, \text{ft}$$

$$\text{plf2by8} := DL2by8 + \text{FloorLoad} \cdot \text{BeamTribWidth} = 202.6 \frac{\text{lbf}}{\text{ft}}$$

$$\text{Moment2by8} := \frac{\text{plf2by8} \cdot L12^2}{8} = 2532.5 \text{lbf} \cdot \text{ft}$$

$$\text{Srequired} := \frac{\text{Moment2by8}}{\sigma\text{allowable}} = 33.767 \text{in}^3$$

$$\text{Sactual2by8} := 13.14 \text{in}^3 \qquad \textbf{No Good!}$$

A redesign is required. Since the architect allowed the use of 10inch beams, a 10inch depth is the logical next step, not the 8inch beam initially chosen.

The next iteration will try a 4by10 beam and the calculations are shown in Fig. 4.6, but other beams that meets the architect's requirements may also be tried.

Fig. 4.6 Subsequent
Project 4–1 calculations

$$\text{DL4by10} := 7.869\frac{\text{lbf}}{\text{ft}}$$

$$\text{plf4by10} := \text{DL4by10} + \text{FloorLoad}\cdot\text{BeamTribWidth} = 207.869\frac{\text{lbf}}{\text{ft}}$$

$$\text{Moment4by10} := \frac{\text{plf4by10}\cdot\text{L12}^2}{8} = 2598.4\text{lbf}\cdot\text{ft}$$

$$\text{Srequired} := \frac{\text{Moment4by10}}{\sigma\text{allowable}} = 34.645\text{in}^3$$

$$\text{Sactual4by10} := 49.9\,\text{in}^3 \qquad \textbf{OK!}$$

$$\sigma\text{actual4by10} := \frac{\text{Moment4by10}}{\text{Sactual4by10}} = 624.732\frac{\text{lbf}}{\text{in}^2}$$

So the shorter beam design is OK, it is not perfectly optimized but that is not the goal of the exercise. The shorter beams on gridlines A and B have less floor load because they are on the edge of the building, thus, they only feel 4ft of floor width (tributary width), but nobody would re-design them as different than the interior beams, so they remain as 4by10 also.

Along the Gridlines B and C, the interior beams will have larger support reactions than will the beams along Gridlines A and D. The calculations are elementary and a qualitative depiction of the different reaction forces is shown in Fig. 4.7.

$$\text{ReactionInteriorBeam} := \frac{\text{plf4by10} \cdot \text{L12}}{2} = 1039.3 \text{lbf}$$

$$\text{plfExteriorBeam} := \text{DL4by10} + \text{FloorLoad} \cdot 4\text{ft} = 107.869 \frac{\text{lbf}}{\text{ft}}$$

$$\text{ReactionExteriorBeam} := \frac{\text{plfExteriorBeam} \cdot \text{L12}}{2} = 539.345 \text{lbf}$$

Fig. 4.7 Reactions from beams become forces on girders

Notice that the interior short beams are supported by the long beams. But the exterior short beams are supported by the columns. Now do the statics on the exterior long beams. Notice that they feel the floor load only as point loads coming in from the beams on Gridlines B and C. The floor load is already accounted for by the shorter beams. Yet of course, they do feel their own self weight, which at this point is still unknown as the beam size has not been assigned. A representation of the beam on Gridline3 is shown in Fig. 4.8.

Fig. 4.8 Representation of beam on gridline 3

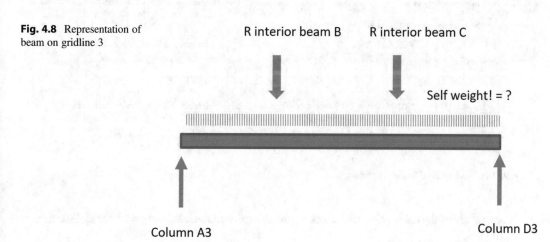

R interior beam B   R interior beam C

Self weight! = ?

Column A3   Column D3

It seems reasonable to guess that the long exterior beams are also 4by10. The symmetry of the problem immediately establishes the fact that the peak moment will be at a cut somewhere between Gridlines B and C. The moment there can be found by statics or by the Müller-Breslau method. The latter is shown in Fig. 4.9. Notice that the loads from the beams are overwhelmingly larger than the self weight of the girder. Both loads are drawn to the same scale to make that point clear.

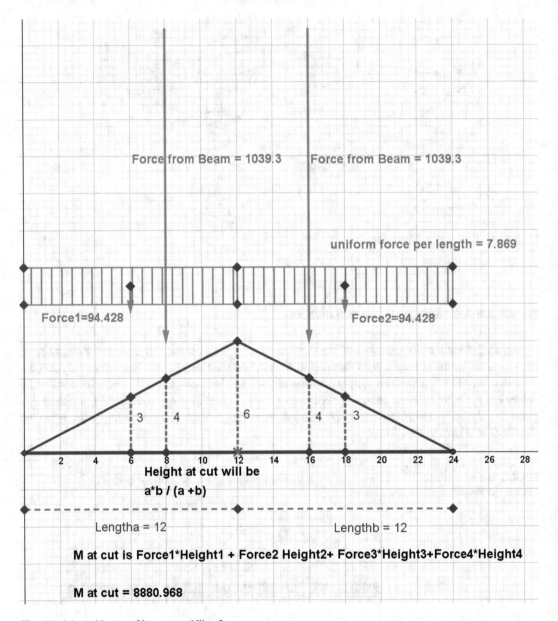

Fig. 4.9 M at mid-span of beam on gridline 3

The summary of the calculations for checking the design, and re-sizing the girder are as described in Fig. 4.10.

Mgirder := 8881.3lbf·ft

$$\sigma\text{actualgirder} := \frac{\text{Mgirder}}{\text{Sactual4by10}} = 2135.36\frac{\text{lbf}}{\text{in}^2}$$

No Good! Stress is over 900

$$\text{Sreqdgirder} := \frac{\text{Mgirder}}{\sigma\text{allowable}} = 118.42\text{in}^3$$

Try a 10by10 beam to maintain a 10in depth

$$\text{DL10by10} := 21.94\frac{\text{lbf}}{\text{ft}}$$

$$\text{S10by10} := 142.9\text{in}^3$$

Mgirder := 9894lbf·ft

$$\sigma\text{actualgirder} := \frac{\text{Mgirder}}{\text{S10by10}} = 830.85\frac{\text{lbf}}{\text{in}^2} \qquad \text{OK!}$$

Fig. 4.10 Beam design calculations

Gravity Loads on Inclined Members

In California, Live Loads are allowed to be projected along a horizontal surface. Dead Load must always be projected along the actual length of the member. The following brief example will highlight the difference. A simply supported beam, shown in Fig. 4.11, has its left end pinned at (0,0) and its right end has a roller at (12,4) in x,z space. The beam is subjected to a vertical, downward 100 force/length Dead Load (DL) along its length.

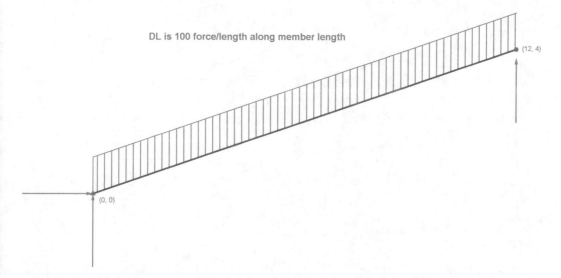

Fig. 4.11 Vertical load on inclined beam

If the same beam is subjected to vertical, downward 100 force/length Live Load (LL) along its horizontal projection, how would that load look? Notice in Fig. 4.12 that the LL along a horizontal projection must be reduced if it is to be equivalent along the longer, inclined length.

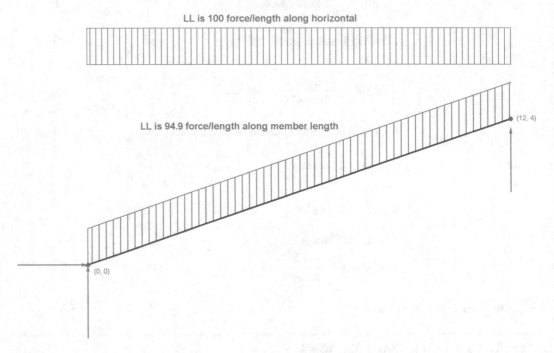

Fig. 4.12 Equivalent horizontal projection of originally inclined load

The calculations to find the equivalent load along the length of beam are quick and they are presented in Fig. 4.13.

$$\text{rise} := 4\text{ft}$$

$$\text{run} := 12\text{ft}$$

$$\text{Length} := \sqrt{\text{rise}^2 + \text{run}^2} = 12.6\text{ft}$$

$$w := 100\frac{\text{lbf}}{\text{ft}}$$

$$\text{Ftot} := w \cdot \text{run} = 1200\text{lbf}$$

$$\text{walongL} := \frac{\text{Ftot}}{\text{Length}} = 94.9\frac{\text{lbf}}{\text{ft}}$$

$$\text{wLiveLoad} := w \cdot \frac{\text{run}}{\text{Length}} = 94.9\frac{\text{lbf}}{\text{ft}}$$

Fig. 4.13 Calculating equivalent horizontal projection load

The Müller-Breslau for reaction, shear or moment for an inclined member subjected to vertical load, is identical to the process for a horizontal member. Remove the effect sought, displace a unit amount, and calculate loft of each load on each folded portion. For instance, the magnitude of the right reaction induced by a vertical point load applied somewhere on an inclined member is found by the construction shown in Fig. 4.14. If a point load is applied downward at $x=6.5$, the right reaction would be as shown in Fig. 4.14.

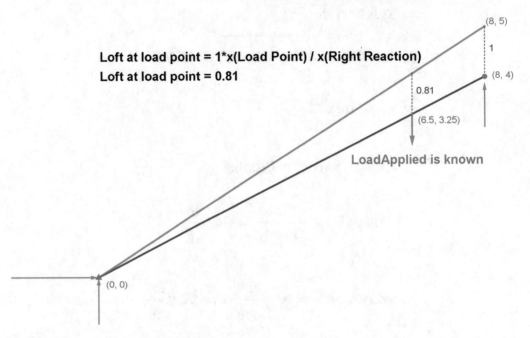

Rright * 1 = Load Applied * Loft at Load Point
Rright = Load Applied * 0.81

Loft at load point = 1*x(Load Point) / x(Right Reaction)
Loft at load point = 0.81

(8, 5)

1

(8, 4)

0.81

(6.5, 3.25)

LoadApplied is known

(0, 0)

Fig. 4.14 Right reaction for inclined beam subject to vertical load

The following inclined beam, shown in Fig. 4.15, is subjected to 2 force/length DL which, as always, is along its actual length. For fun, place that DL along a horizontal and check:

$$M_{midspan} = \frac{w_{projected} \cdot run^2}{8}$$

(4.4)

Compare the $M_{midspan}$ from Eq. (4.4) to the Müller-Breslau approach for moment at mid-span. Also find the magnitude of the right reaction and the magnitude of the shear at mid-span. The problem is shown in Fig. 4.15.

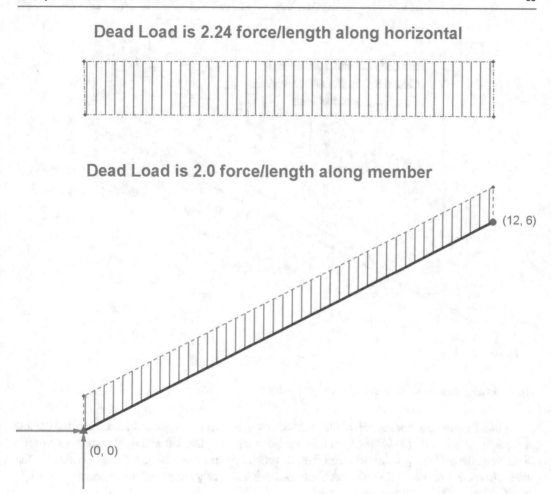

Fig. 4.15 Example problem of inclined beam

To find the right reaction, loft up the right reaction 1 unit of length. Find the centroid of the entire load as there are no folds in the lofted structure. Find the loft at the position of the centroid of the load. The right reaction is that centroidal force multiplied by the loft. This is shown in Fig. 4.16.

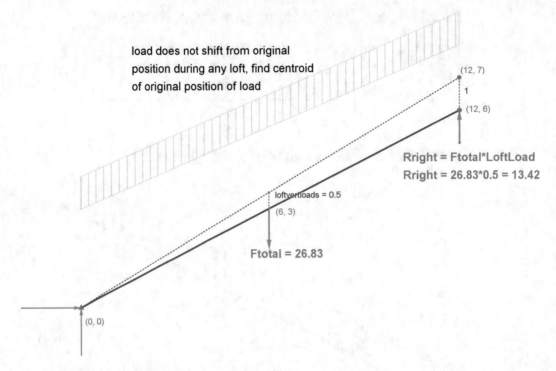

load does not shift from original
position during any loft, find centroid
of original position of load

(12, 7)

1

(12, 6)

Right = Ftotal*LoftLoad
Right = 26.83*0.5 = 13.42

loftvertloads = 0.5
(6, 3)

Ftotal = 26.83

(0, 0)

Fig. 4.16 Right reaction for inclined beam example problem

To find the moment at mid-span of the inclined beam, use the technique as before, first find the loft at the kink, this is $a \cdot b / (a+b)$ where a and b are horizontal lengths, left and right respectively of the kink. Now, the uniformly distributed load must be broken up into two sub-loads as there is a fold in the lofted structure. The loft in line with each sub load is immediately found and the moment at the cut is found as before. This is shown in Fig. 4.17.

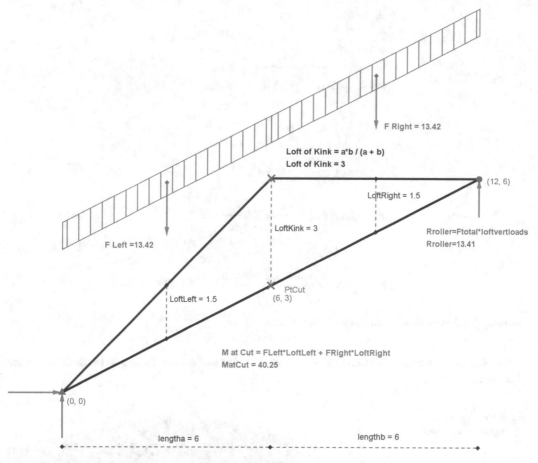

Fig. 4.17 Moment at mid-span for inclined beam example problem

As was shown in Fig. 4.17, the Müller-Breslau method of finding moment Mmidspan=40.25 force·length exactly matches the easy algebraic approach based on horizontal spans. The algebraic approach is shown in Fig. 4.18.

$$\text{rise} := 6\text{ft}$$

$$\text{run} := 12\text{ft}$$

$$\text{Length} := \sqrt{\text{rise}^2 + \text{run}^2} = 13.4\text{ft}$$

$$w := 2\,\frac{\text{lbf}}{\text{ft}}$$

$$\text{Ftot} := w \cdot \text{Length} = 26.8\text{lbf}$$

$$\text{walongHoriz} := \frac{\text{Ftot}}{\text{run}} = 2.2\,\frac{\text{lbf}}{\text{ft}}$$

$$\text{Mmidspan} := \frac{\text{walongHoriz} \cdot \text{run}^2}{8} = 40.2\text{lbf} \cdot \text{ft}$$

Fig. 4.18 Verification of moment calculation using algebra

Suppose the roof beams of a building were closely spaced as shown in the Fig. 4.19. Roof beams are also known as "rafters".

Fig. 4.19 Graphical representation of rafters

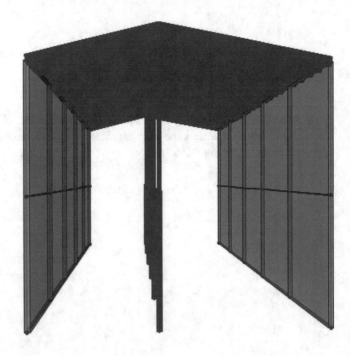

The roof beams ("rafters") rest on bearing walls. The roof beams have a spacing that is small, typically something like 16inches "on center" (o.c.). Each roof beam bends, but what does the bearing wall feel? At very close spacing, the load experienced by the support walls can be assumed to be a uniformly distributed load, not a lot of very closely spaced point loads.

Now look at these roof beams (rafters) more closely. Suppose the slope of each side of the roof is known. A roof (downward) load of **force/area** is applied uniformly onto the roof.

How to change **force/area** to **force/length** on an inclined member? Remember that the inclined members have a known spacing between them.

Step 1 is to recognize the transformation from force/area to force/length

$$\frac{force}{L^2} \cdot L = \frac{force}{L}$$ (4.5)

Therefore:

$$\frac{force}{Length} = \frac{force}{area} \cdot spacing$$ (4.6)

Step 2 is to note whether the downward load is applied on a horizontal projection as LL is, or per actual length as DL is.

Snow load can be assumed to be applied on a horizontal projection, but dead load never can. Dead load must be applied per actual length. Consider the building structure shown in Fig. 4.20. Closely spaced rafters meet at the peak point of the roof. Rafters rest on bearing walls.

Fig. 4.20 Closely spaced rafters

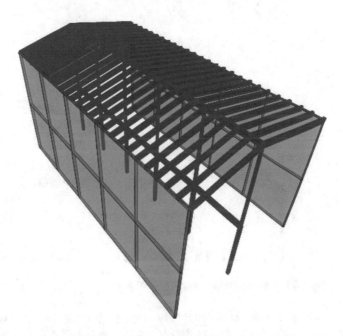

For the building shown in Fig. 4.20, the rafters have the following properties:

- 4 to 12 slope
- 14 unit horizontal span of the rafters
- roof beams (rafters) was spaced 2 units of length on center

- snow load 30 force/area
- dead load 18 force/area which accounted for the rafters themselves plus insulation.
- Bearing walls weigh 10 force/area and are 16 units of length tall

The Müller-Breslau method of finding the bending moment at the mid-span of a typical rafter, is shown in Fig. 4.21. The fact that there is both LL and DL and that they are projected along varying surfaces poses no additional difficulty with the Müller-Breslau method.

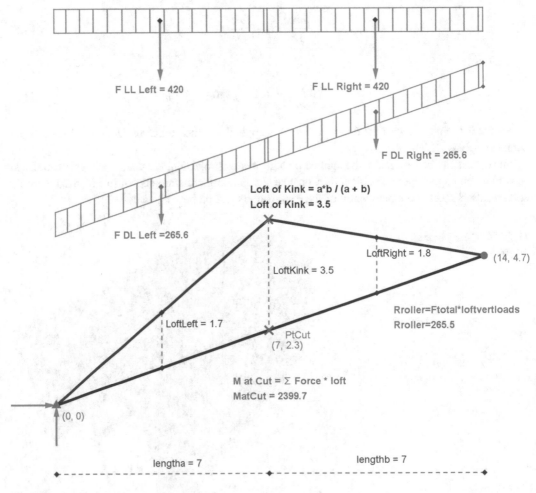

Fig. 4.21 Bending moment at mid-span of rafter

The next step is to find the force along the top of the bearing wall that arises from the rafters. The Müller-Breslau method of finding the left reaction of a typical rafter, i.e. the force at the top of the wall due to one rafter, is shown in Fig. 4.22. Notice how the width of the bearing wall never comes into play.

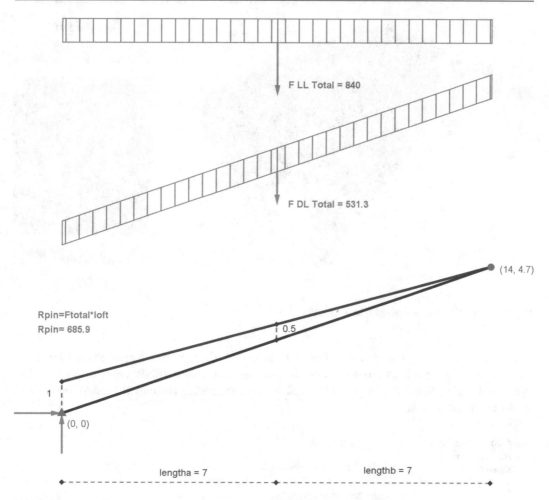

Fig. 4.22 Left reaction of rafter

Figure 4.22 showed that the force needed to hold up one rafter at the top of the bearing wall is 686 units of force. The force of one single rafter support reaction at the top of the wall is reversed and applied downward to the top of the bearing wall. Furthermore, this single reaction force is changed to a force/length along the top of the wall via the spacing of the rafters. A dimensional analysis will certainly help here. If the spacing was 24 inches (2 ft) and the force was 686lb then

$$Uniform\ Load = \frac{686\ force}{2\ length} = 383\ {}^{force}\!/_{length} \tag{4.7}$$

Next, look at the bearing walls. Clearly the top of the wall has a uniform load, but what happens at the bottom of the wall? An 3D view and an elevation view of the wall is shown in Fig. 4.23.

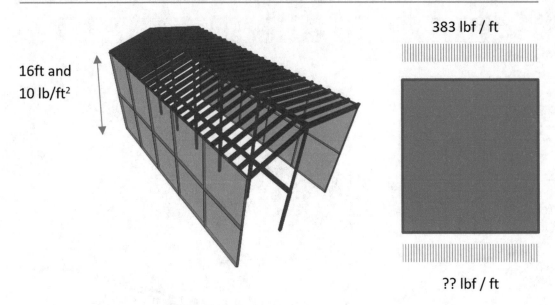

16ft and
10 lb/ft²

383 lbf / ft

?? lbf / ft

Fig. 4.23 Graphical representation of bearing wall

Notice that wall weight is typically given in force/length². Change force/length² to force/length by multiplying by the height of the wall! Since the bearing wall weighs 10lb/ft² and the height of this upper wall is 16ft, then at the bottom of the wall there is a supporting reaction of 543lb/ft. Figure 4.24 shows the quick calculation.

Fig. 4.24 Calculation of reaction at base of bearing wall

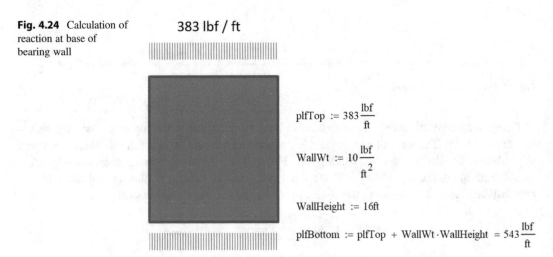

383 lbf / ft

$$plfTop := 383 \frac{lbf}{ft}$$

$$WallWt := 10 \frac{lbf}{ft^2}$$

$$WallHeight := 16ft$$

$$plfBottom := plfTop + WallWt \cdot WallHeight = 543 \frac{lbf}{ft}$$

Project 4–2 Task Analyze a bearing wall holding up roof rafters. The roof is subject to a snow load which is projected on a horizontal plane. The Dead Load applies to the roof rafters and it accounts for the weight of the rafters themselves. The roof rafters rest on bearing walls that have some known force/area that is reasonable.

You must:

- Present a clear set of drawings (either by hand or by computer) with clear dimensions and presentation of assumed loads.
- Do not use a 4:12 slope, use something else
- Present a span of the roof rafters that is not 14, use something else
- Calculate the loads at the ends of the rafters, at the top of the bearing walls and at the bottom of the bearing walls.
- You must check your work. Convince yourself that the statics check out correctly! Show this check convincingly.
- Clear, neat calculations with units at the end of the calculations. These can be done on the computer or by hand.
- Peak bending moment of the roof rafter with units

Sample student work for Project 4–2 is shown at the end of this chapter.

Live Loads

Live load is the load superimposed by the use and occupancy of the building, not including wind load, earthquake load or dead load. It includes the weight of people, furniture etc. Live load varies as people move around or as furniture is rearranged. Although live load is essentially dynamic (i.e. time varying) we usually assume it to be static and vertical (downward) load.

The minimum live loads required by the IBC are found in Table 1607.1. These loads apply to roof and to floor live loads. Floors in buildings where partition locations are subject to change must be designed to support a load of 15 lb/ft^2 in addition to all other loads.

Should we design every structure for every conceivable maximum live load combination? The philosophy behind this idea is that if the loads are not too great (ASCE 4.7.3) and if the area considered is fairly large (ASCE 4.7.2) then a conservative reduction (i.e. not too great) is allowed for Live Load.

The *International Building Code* (IBC) has finally been adopted in California. It generally absorbs other codes into it, for example *ASCE 7, **Minimum** Design Loads for Buildings and Other Structures* is a detailed description of loads, and this code has been adopted or absorbed by IBC. Material specific codes have also been adopted by IBC such as the ACI 318 for concrete, AISC code for steel. A key idea in all building codes is the word minimum. It is the legal minimum that must be adhered to.

Live loads can be estimated, but final minimum values must ultimately checked via the local building code. Live loads caused by the building's contents or objects are often called occupancy loads. Roof live loads are often differently handled because they are induced by water accumulation, snow, etc. Floor live loads include the weights of people, furniture, books, etc. and other features not included in the dead load. Figure 4.25 exhibits typical live loads.

Typical floor live loads	psf
Residential	40
Offices	50 to 80
Stores	75
Partition loads and minimum roof loads	20
Balconies < 100 ft^2	80
Corridors	80
Stairs	100
Public assembly areas	100

Note: 100ft^2 = 9.3m^2

Fig. 4.25 Typical live loads

From the typical values of the Fig. 4.25, it is apparent that public areas such as corridors must carry more live load than living or working areas. Also, office buildings weigh more than apartment buildings. These ideas simply must have design implications.

It is improbable that the live load in a building will fully cover a large tributary area to the same extent as a small area. For example, it is not likely that, in a multi-story structure, every floor simultaneously will carry the full live load. Building codes take this into account by allowing the use of floor live load reduction factors, when the factored tributary area supported by a structural member is 400 ft^2 or more (ASCE 4.7.2) and the live load does not exceed 100 psf. (ASCE 4.7.3)

$$L = L_o \left(0.25 + \frac{15}{\sqrt{K_{LL} \cdot A_T}} \right) \tag{4.8}$$

Where L = reduced LL, L_o = unreduced LL, K_{LL} = LL factor (found in ASCE Table 4.2) and A_T = tributary area (ft^2).

Fig. 4.26 shows how K_{LL} is found.

Element	K_{LL}[a]
Interior columns	4
Exterior columns without cantilever slabs	4
Edge columns with cantilever slabs	3
Corner columns with cantilever slabs	2
Edge beams without cantilever slabs	2
Interior beams	2
All other members not identified, including: 　Edge beams with cantilever slabs 　Cantilever beams 　One-way slabs 　Two-way slabs 　Members without provisions for continuous shear transfer normal to 　　their span	1

[a]In lieu of the preceding values, K_{LL} is permitted to be calculated.

Fig. 4.26 Live load element factor K_{LL}

Exceptions exist! The reduction cannot end up with less than 50% of the original live load (one floor loading). And roofs experience yet a different formula for live load reductions.

Now look at the previous Eq. (4.8) which is 4.7–1 in ASCE 7–10 for the new live load L. Notice that the formula is valid only for the case where $K_{LL}A_T$ is greater than 400ft². At $K_{LL}=1$ and $A_T=400$ft², the bracketed term becomes $0.25 + 0.75 = 1$, no reduction.

$$L = L_o\left(0.25 + \frac{15}{\sqrt{1 \cdot 400}}\right) \tag{4.9}$$

$K_{LL}A_T$ of ASCE Eq. 4.7.3 is an interesting and important idea. It captures influence area as opposed to tributary area. The difference between these is explained in Fig. C4–1 on page 413 of the ASCE Code. This is the live load commentary. It is shown in Fig. 4.27.

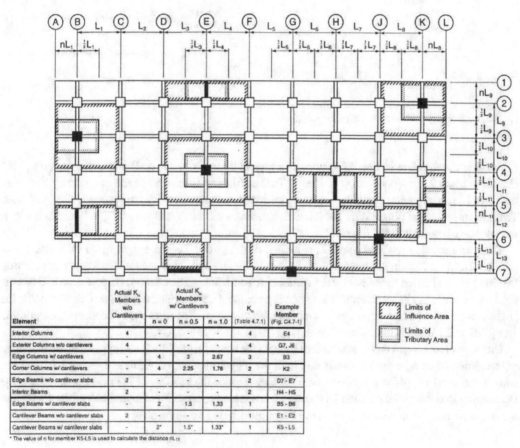

Element	Actual K_{LL} Members w/o Cantilevers	Actual K_{LL} Members w/ Cantilevers			K_{LL} (Table 4.7.1)	Example Member (Fig. C4.7-1)
		n = 0	n = 0.5	n = 1.0		
Interior Columns	4	-	-	-	4	E4
Exterior Columns w/o cantilevers	4	-	-	-	4	G7, J6
Edge Columns w/ cantilevers	-	4	3	2.67	3	B3
Corner Columns w/ cantilevers	-	4	2.25	1.78	2	K2
Edge Beams w/o cantilever slabs	2	-	-	-	2	D7 - E7
Interior Beams	2	-	-	-	2	H4 - H5
Edge Beams w/ cantilever slabs	-	2	1.5	1.33	1	B5 - B6
Cantilever Beams w/o cantilever slabs	2	-	-	-	1	E1 - E2
Cantilever Beams w/ cantilever slabs	-	2*	1.5*	1.33*	1	K5 - L5

* The value of n for member K5-L5 is used to calculate the distance nL_{11}

FIGURE C4.7-1 Typical Tributary and Influence Areas

436   STANDARD ASCE/SEI 7-16

Fig. 4.27 Tributary area versus influence area

The following example shown in Fig. 4.28 examines live load reduction.

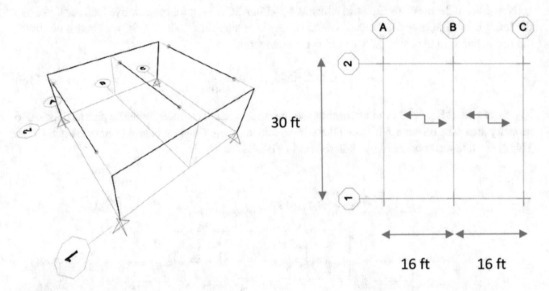

Fig. 4.28 Geometry for live load reduction example

Assume the structure in Fig. 4.28 supports a typical floor. Gridlines A, B and C are spaced at 16ft. Gridlines 1 and 2 are spaced at 30 ft. Heavy Timber framing is used to create the primary structure, with 2×12 joists running East/West between gridlines A, B and C which are spaced 1.33ft. on center. These joists feel 10psf (this is lb/ft^2) of DL which includes the joists' self weight. The floor LL is 40psf. Include the LL reduction for a typical floor.

There is a subtlety related to the LL reduction of the girder (i.e. the bigger beam on gridline 1 or gridline 2). LL reduction is now calculated element by element. That means that simply transferring live load from element to element via traditional load tracing statics cannot be performed. However dead load can always be transferred this way, via traditional statics. For the beam on Gridline B, the LL got reduced because of the large area associated with this beam. But what about the element on gridline 1 or gridline 2? What is its LL reduction?

One way of looking at this is to take the DL from the beam on gridline B and apply it at the mid-span of the girder. Then take the LL associated with the bigger element on gridline 2, using KLL =2 but what is the area? Fig. 4.29 shows one such LL area. Notice the subtlety of the previous sentence, the DL flowing into the girder on Line 2 is based on static load tracing. The LL that that girder feels from the connection at B2 flows into it from the beam on B based on some area.

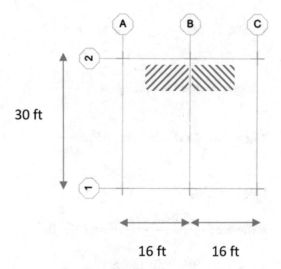

Fig. 4.29 Live load flowing from beam into girder at B2

Another way of looking at this is to say KLL = 4! Reasoning is that if the girder fails, the entire floor goes. The code allows you to make such judgments. It is acceptable to use either KLL for the element on gridline 2, but it is wrong to simply transfer the LL from the Beam on B statically.

The calculations for DL and LL flow follow in Fig. 4.30. Start with the Joists, but no Live Load reduction for the joists is expected since their area is very small.

$$L12 := 30 \text{ft}$$

$$LAB := 16 \text{ft}$$

$$LBC := 16 \text{ft}$$

$$\text{spacing} := 1.33 \text{ft}$$

$$\text{Ajoist} := \text{spacing} \cdot LAB = 21.28 \text{ft}^2$$

$$KLL := 2$$

$$A := \text{Ajoist} \cdot KLL = 42.56 \text{ft}^2$$

Fig. 4.30 Live load reduction calculations on joists

As expected, this is less than 400ft^2, so no LL reduction is allowed.

Next, find the DL and the LL on joists. These steps are shown in Fig. 4.31.

$$\text{floorDL} := 10\frac{\text{lbf}}{\text{ft}^2} \qquad \text{floorLL} := 40\frac{\text{lbf}}{\text{ft}^2}$$

$$\text{wDLjoist} := \text{floorDL}\cdot\text{spacing} = 13.3\frac{\text{lbf}}{\text{ft}}$$

$$\text{wLLjoist} := \text{floorLL}\cdot\text{spacing} = 53.2\cdot\frac{\text{lbf}}{\text{ft}}$$

$$\text{Ljoist} := \text{LAB}$$

$$\text{RDLjoist} := \frac{\text{wDLjoist}\cdot\text{Ljoist}}{2} = 106.4\,\text{lbf}$$

$$\text{RLLjoist} := \frac{\text{wLLjoist}\cdot\text{Ljoist}}{2} = 425.6\,\text{lbf}$$

$$\text{RDLLLjoist} := \text{RDLjoist} + \text{RLLjoist} = 532\,\text{lbf}$$

Fig. 4.31 Dead load and live load calculations on joists

Now do the beam analysis on Gridline A or C with the LL reduction. The load going to the element on Gridline A is shown in Fig. 4.32.

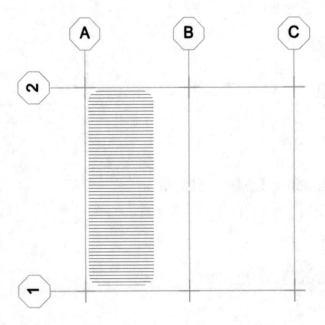

$$\text{AreaBeamA} := \text{L12}\cdot\frac{\text{LAB}}{2} = 240\,\text{ft}^2$$

$$\text{KLL} := 2$$

$$\text{KLLAT} := \text{KLL}\cdot\text{AreaBeamA} = 480\,\text{ft}^2$$

Fig. 4.32 Load flowing into beam on gridline A

A reduction is allowed! Note the Code uses Lo for Load Original. Sorry this is not a length! Fig. 4.33 described the new LL.

Fig. 4.33 Live load reduction for beam on gridline A

$$Lo := \frac{floorLL}{\frac{lbf}{ft^2}} = 40 \qquad\qquad KLLAT := \frac{KLLAT}{ft^2} = 480$$

$$Lnew := Lo \cdot \left(0.25 + \frac{15}{\sqrt{KLLAT}} \right) \cdot \frac{lbf}{ft^2} = 37.386 \frac{lbf}{ft^2}$$

Thus what LL force/length does beam A feel? The tributary width is 8ft and the new LL was just calculated as force/length^2.

$$wLLbeamA = Lnew \cdot \frac{LAB}{2} = 299 \; ^{lb}/_{ft} \tag{4.10}$$

Before finding the final force/length on beam A, the Dead Load must be changed from force/length^2. This will be done two different ways to create a nimbleness in the approaches used for load flow. The first way is the immediately obvious way of doing this, namely to take the DL of 10 force/length^2 and to multiply that by the tributary width associated with the beam on Gridline A.

$$wDLbeamAversion1 = floorDL \cdot \frac{LAB}{2} = 80 \; ^{lb}/_{ft} \tag{4.11}$$

The second version of this calculation is a bit more subtle. It will take the DL reactions (force) from the joists and expands them to become force/length via the spacing of the joists framing into the beam on A.

$$wDLbeamAversion2 = \frac{RDLjoist}{spacing} = 80 \; ^{lb}/_{ft} \tag{4.12}$$

Beam A could now be designed assuming both DL and LL are uniformly distributed because there are many, many LL point loads hitting beam A from the joists. There is no benefit from analyzing this beam as being subject to point loads. DL and LL can be factored separately using Code prescribed factors for each, if load factors are called for. In allowable stress design for timber, and for deflection analysis for timber, no factoring of loads is used. If the allowable stress is known for heavy timber is known, say 900 lb/in^2, then an appropriate section modulus can be found as in Fig. 4.34.

Fig. 4.34 Design of beam on gridline A

$$wDLbeamA := wDLbeamAversion1$$

$$MmaxA := \frac{(wDLbeamA + wLLbeamA) \cdot L12^2}{8} = 42647.5 \, lbf \cdot ft$$

$$\sigma allowed := 900 \frac{lbf}{in^2}$$

$$SrequiredA := \frac{MmaxA}{\sigma allowed} = 568.6 \, in^3$$

Now look at Dead Load of Beam B. Notice that the joists hit it from the east and from the west. Again, there are two ways of finding the DL force/length on beam B. The first is immediately obvious, namely to take the DL force/length^2 and to multiply that value by the tributary width associated with beam B. The second is to expand a force through the spacing. Both approaches are shown in Fig. 4.35.

Fig. 4.35 Dead load on beam on gridline B two different ways

$$\text{wDLbeamBversion1} := \text{floorDL} \cdot \left(\frac{\text{LAB}}{2} + \frac{\text{LBC}}{2} \right) = 160 \frac{\text{lbf}}{\text{ft}}$$

$$\text{wDLbeamBversion2} := \frac{\text{RDLjoist}}{\text{spacing}} \cdot 2 = 160 \frac{\text{lbf}}{\text{ft}}$$

Now look at Live Load on Beam B. Note, statics on the entire grid of beams cannot be used to transfer LL from one element to another because the magnitude of the LL is being reduced during each step of the calculation. Figure 4.36 shows the steps needed.

Fig. 4.36 Live load reduction on beam on gridline B

$$\text{AreaBeamB} := \text{L12} \cdot \left(\frac{\text{LAB}}{2} + \frac{\text{LBC}}{2} \right) = 480\text{ft}^2$$

$$\text{KLL} := 2$$

$$\text{KLLAT} := \text{KLL} \cdot \text{AreaBeamB} = 960\text{ft}^2$$

LL reduction is allowed here.

$$\text{Lo} := \frac{\text{floorLL}}{\dfrac{\text{lbf}}{\text{ft}^2}} = 40 \qquad\qquad \text{KLLAT} := \frac{\text{KLLAT}}{\text{ft}^2} = 960$$

$$\text{Lnew} := \text{Lo} \cdot \left(0.25 + \frac{15}{\sqrt{\text{KLLAT}}} \right) \cdot \frac{\text{lbf}}{\text{ft}^2} = 29.4 \frac{\text{lbf}}{\text{ft}^2}$$

Be careful to ensure that the new load cannot be reduced to less than 50% of the original load. Here it is not lower than that limit, so everything is OK.

$$LLbeamB = LLnew = 29.4 \; {^{lb}/_{ft}} \tag{4.13}$$

Change this Live Load with units of force/length^2 to a force/length now.

$$wLLbeamB = LLbeamB \cdot \left(\frac{LAB}{2} + \frac{LBC}{2} \right) = 469.8 \; {^{lb}/_{ft}} \tag{4.14}$$

Calculate the worst bending moment on BeamB. This is of course due to both DL and LL on BeamB. The elementary design calculations are shown in Fig. 4.37

Fig. 4.37 Design of beam
on gridline B

$$\text{wDLbeamB} := \text{wDLbeamBversion1}$$

$$\text{MmaxB} := \frac{(\text{wDLbeamB} + \text{wLLbeamB}) \cdot \text{L12}^2}{8} = 70856.9 \text{lbf} \cdot \text{ft}$$

$$\sigma\text{allowed} := 900 \frac{\text{lbf}}{\text{in}^2}$$

$$\text{SrequiredB} := \frac{\text{MmaxB}}{\sigma\text{allowed}} = 944.8 \, \text{in}^3$$

To calculate the load flowing into beam2, once again consider the DL separately from the LL. The DL can be followed via statics, i.e. the DL reactions of beamB connect into beam2. But the LL does not work that way.

$$ReactionDLbeamB = \frac{wDLbeamB \cdot L12}{2} = 2400 \ lb \tag{4.15}$$

For LL, what passes through the connection from beamB to the beam2 must be consider, not by statics but by a geometric analysis. This will calculate the new LL on the beam 2 that arises from the point connection with beamB. An instructive way of looking at this area is to think of one half of the live load that acts on beamB, that must pass through the connection to beam2. This is shown in Fig. 4.38. Other loads assigned to beamA and beamC do not flow through this connection. It may be tempting to think of this load as simply the static reaction of half of the load applied to beam2. But that would be incorrect because the load on beam2 was derived for DL on beam2 and LL on beam2 which had its own reduction. A new reduction is applied to the live load passing through the connection from beamB to beam2. Figure 4.38 shows that load. The mild paradox here is that to study the live load on beam2, the live load area acting on beamB which flows through the connection between beamB and beam2 must be found. Yet no statics will be performed to find that live load.

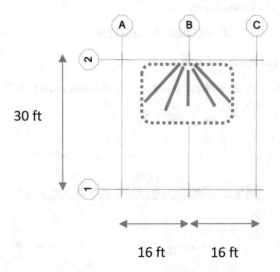

Fig. 4.38 Load flowing into connection at B2

Figure 4.39 shows the steps taken to establish the new LL.

Fig. 4.39 Live load
reduction flowing
through connection
at B2

$$\text{AreaBeamBflowing} := \frac{1}{2} \text{L12} \cdot \left(\frac{\text{LAB}}{2} + \frac{\text{LBC}}{2} \right) = 240 \text{ft}^2$$

$$\text{KLLAT} := \text{KLL} \cdot \text{AreaBeamBflowing} = 480 \text{ft}^2$$

$$\text{Lo} := \frac{\text{floorLL}}{\dfrac{\text{lbf}}{\text{ft}^2}} = 40 \qquad\qquad \text{KLLAT} := \frac{\text{KLLAT}}{\text{ft}^2} = 480$$

$$\text{Lnew} := \text{Lo} \cdot \left(0.25 + \frac{15}{\sqrt{\text{KLLAT}}} \right) \cdot \frac{\text{lbf}}{\text{ft}^2} = 37.4 \frac{\text{lbf}}{\text{ft}^2}$$

$$\text{LLbeamB} := \text{Lnew} = 37.4 \frac{\text{lbf}}{\text{ft}^2}$$

This is not less than 50% of the original, so OK! Thus the new LL on beamB is 37.4lb/ft^2. The LL force passing through this connection is:

$$\text{PointLoadLL} = \text{LLbeamB} \cdot \text{AreaBeamBflowing} = 8973 \; lb \tag{4.16}$$

Then the load passed to 2 is:

$$\text{PointLoadBto2} = \text{ReactionDLbeamB} + \text{PointLoadLL} = 11373 \; lb \tag{4.17}$$

This load on beam2 is summarized in Fig. 4.40.

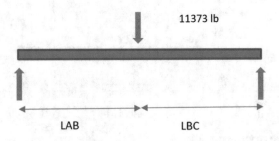

Fig. 4.40 Graphical representation of load on beam on gridline 2

The peak bending from this point load is not $wL^2/8$, it is $PL/4$. The moment and section modulus calculations are shown in Fig. 4.41.

Fig. 4.41 Design of beam on gridline 2

$$\text{MpeakBeam2} := \frac{\text{PointLoadBto2} \cdot (\text{LAB} + \text{LBC})}{4} = 90981.4\,\text{ft} \cdot \text{lbf}$$

$$\text{Sreqdbeam2} := \frac{\text{MpeakBeam2}}{\sigma\text{allowed}} = 1213.1 \cdot \text{in}^3$$

As a final thought, the question of "what is the tributary area that flows into the Beam on 2"? is answered via the extremely powerful Müller-Breslau method. The premise here is now to loft the shear connection from the beam on B to the beam on 2, but to apply a unit load of 1lb/ft^2 on the floor. The answer will be lb on the connection, but that answer really means ft^2, because it presumes we know the load as 1lb/ft^2, but we do not know the area! Fig. 4.42 shows the construction,. The answer is 240ft^2.

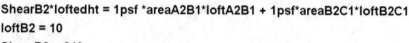

ShearB2*loftedht = 1psf *areaA2B1*loftA2B1 + 1psf*areaB2C1*loftB2C1
loftB2 = 10
ShearB2 = 240

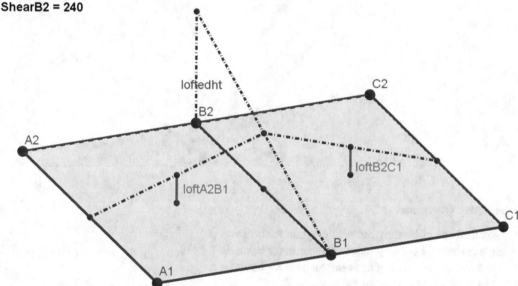

Fig. 4.42 Finding tributary area via Müller-Breslau

In Sect. 4.7 of ASCE 7, the rules for LL reduction of floors are shown. The most important rule to understand is that load tracing can be done for DL, but LL is constantly changing from element to element and the tributary area is the area that flows into the higher order structural element. Using statics is tedious, but it can be done. To use statics to find the tributary area, apply 1lb/ft^2 on the entire floor. Then, eventually after tracing the load to the connection of interest, or to the element of interest consider the following:

$$\frac{connection\ load\ from\ 1\ \frac{lb}{ft^2}}{entire\ floor\ area} = \frac{connection\ load\ from\ LL\ \frac{lb}{ft^2}}{tributary\ area} \tag{4.18}$$

Important There is only ONE tributary area for each item being considered. That area experiences the original dead load and it also experiences the reduced live load.

Project

Another example, which could be used as a project, is shown in Fig. 4.43. Answer the qualitative questions first. Then find the moment mid-span between 2 and 3 on the beam along A. DL includes self-weight of all beams and joists is DL=50lb/ft^2, LL=60lb/ft^2.

Fig. 4.43 Quantitative example building geometry and spacing

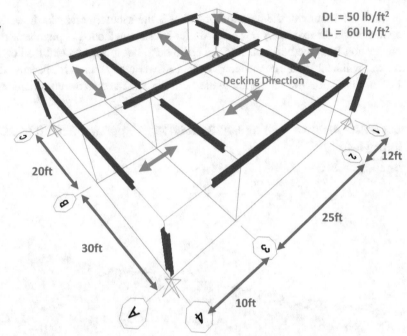

DL = 50 lb/ft^2
LL = 60 lb/ft^2

Decking Direction

20ft

12ft

25ft

30ft

10ft

Qualitative Questions

- Sketch shape of bending moment diagram for beam on A
- Sketch shape of bending moment diagram for beam on 1
- Sketch shape of bending moment diagram for beam on C
- Sketch the loading diagram for beam on B

Quantitative Questions

- Find the bending moment midway between lines 2 and 3 for the beam on A

The biggest concern for the quantitative question is what is the tributary area for each of the point loads connecting into the beam on A. If that question is correctly answered, the LL is reduced based on that area and the magnitude of the DL point loads and the LL point loads are immediately found.

To answer this via statics, one would apply 1lb/ft^2 everywhere and trace the load to the connection at A2 and to the connection at A3. The bad news is that this is tedious. The good news is that if done correctly, then each of those areas can be individually used for DL and for the individually reduced LL. Notice, the LL reduction may well differ for the connection at A2 than for the connection at A3. This is exactly why LL reduction bewilders some students. It must be done on each point of load flow into the higher order element.

The answer to this via Müller-Breslau is insightful, fast and nearly foolproof. And of course, it can be done on a table napkin sketch! Fig. 4.44 shows the tributary area for the connection at A2.

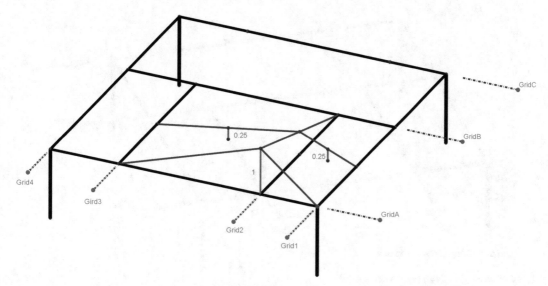

$$TribA2 = Area12AB * 1lb/ft^2 * 0.25 + Area23AB * 1lb/ft^2 * 0.25 = 262.5ft^2$$

Fig. 4.44 Tributary area for the connection at A2

Figure 4.45 shows the tributary area for the connection at A3.

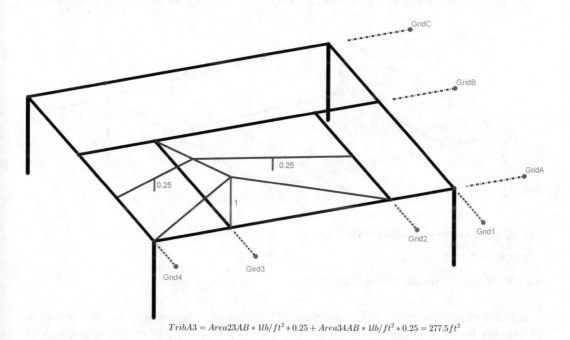

$$TribA3 = Area23AB * 1lb/ft^2 * 0.25 + Area34AB * 1lb/ft^2 * 0.25 = 277.5ft^2$$

Fig. 4.45 Tributary area for the connection at A3

Figure 4.46 shows the DL and LL flowing into the beam on Gridline A. Notice the unique reductions in live load for each of the connections.

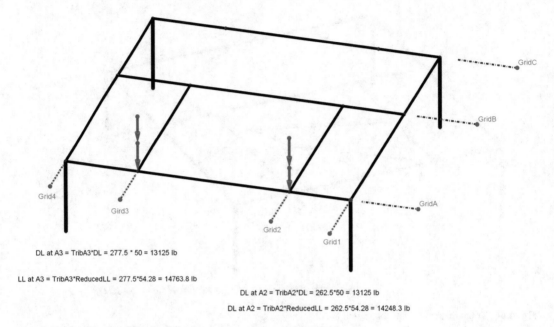

DL at A3 = TribA3*DL = 277.5 * 50 = 13125 lb

LL at A3 = TribA3*ReducedLL = 277.5*54.28 = 14763.8 lb

DL at A2 = TribA2*DL = 262.5*50 = 13125 lb

DL at A2 = TribA2*ReducedLL = 262.5*54.28 = 14248.3 lb

Fig. 4.46 Dead load and live load flowing into connection at A2 and A3

Figure 4.47 shows the process to find the moment at the mid-span point of the beam on Gridline A.

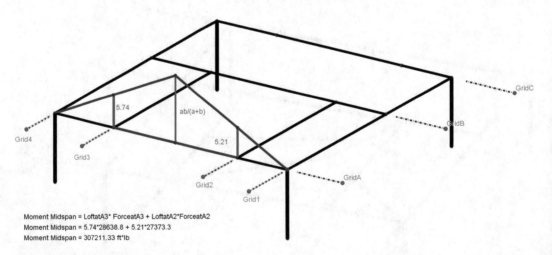

Moment Midspan = LoftatA3* ForceatA3 + LoftatA2*ForceatA2
Moment Midspan = 5.74*28638.8 + 5.21*27373.3
Moment Midspan = 307211.33 ft*lb

Fig. 4.47 Moment at mid-span of beam on gridline A

Students were given Project 4–2 to complete, and the following figures show sample portions of their reports. Figures 4.48 and 4.49 show sample output comparing SAP2000 to hand calculations via MATLAB.

```
format compact
%Symbolic Tool Kit
syms lb ft
%Define Variables
snowpsf_horiz = 23*lb/ft^2;          %snow load
deadpsf_inc = 33*lb/ft^2;            %dead load
wall_dead = 27*lb/ft^2;              %wall dead load
wall_height = 20*ft;                 %wall height
spacing = 0.5*ft;                    %rafter spacing
rafterspan_horiz = 16*ft;            %horizontal rafter span
rise = 5;                            %roof rise
run = 12;                            %roof run
%Slope Calculations
hyp = sqrt(rise^2 + run^2);          %hypotenuse
sfactor_big = hyp/run;               %slope factor for inflation
sfactor_small = run/hyp;             %slope factor for deflation
%PLF Transformations
snowplf_horiz = snowpsf_horiz*spacing;
deadplf_inc = deadpsf_inc*spacing;
wallplf = wall_dead*wall_height;                %wallplf
%Slope Transformations
snowplf_inc = snowplf_horiz*sfactor_small;
deadplf_horiz = deadplf_inc*sfactor_big;
%Totals
horiz_loadplf = deadplf_horiz + snowplf_horiz;  %total horizontal load
inc_loadplf = deadplf_inc + snowplf_inc;        %total inclined load
disp('The total inclined load is:')
vpa(inc_loadplf, 4)
%Rafter Reactions
reac_wall = horiz_loadplf*rafterspan_horiz/2;
disp('The reaction on the wall is:')
vpa(reac_wall, 4)
rafterwallplf = reac_wall/spacing;
disp(rafterwallplf)
%Wall Reactions
reac_bottom = rafterwallplf + wallplf;
disp('The reaction at the bottom of the wall is:')
vpa(reac_bottom, 4)
```

Fig. 4.48 Student sample work MATLAB code Project 4–2

36 unit wide, 12 unit tall wall with two door openings
Dead Load analysis of qualitative load flow

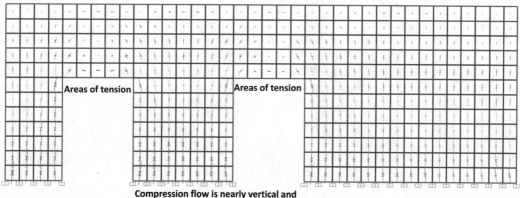

Areas of tension   Areas of tension

Compression flow is nearly vertical and
an order of magnitude greater than tension stress

Fig. 4.49 Student sample work, qualitative load flow Project 4–2

Earthquake Loads

5

The Nature of Earthquakes

Earthquakes are greatly feared because tremendous forces can be unleashed with little or no warning. If strong earthquakes last more than several seconds, they can cause widespread destruction. Fortunately, most earthquakes are not too strong, do not last too long, and consequently cause little or no permanent damage to well designed structures. Hooray for architectural engineers!

The theory that best explains earthquakes is that the earth's crust is made up of huge rock plates floating on a layer of molten rock. As these plates move, friction and other forces lock them together at certain regions. As the strain increases at these junctures over long periods of time, the frictional force between the plates is finally overcome and the plates violently slip past each other to a new unstrained position. This slip releases complex shock waves, some waves are up/down, others are back/forth, still others are rolling. In 10–20 s, the worst part of the earthquake is usually over, though aftershocks can still occur. This theory focuses on the faults, or boundaries between adjacent plates, such as the San Andreas fault in California. Note that strong earthquakes may result in vertical or lateral shifts anywhere from a few inches to 20 feet! The location in the earth's crust where the rock slippage begins is called the focus, or the hypocenter of the earthquake. Its depth is usually 5–10 miles in California. The projection of the focus on the ground surface is called the epicenter.

Measuring Earthquakes

Earthquakes are measured by two distinctly different scales: the Richter scale and the Modified Mercalli scale. The Richter scale is a measure of the earthquake's magnitude, i.e. the amount of energy it releases. The scale is logarithmic, with each number representing roughly 33 times the energy of the preceding number. Thus a magnitude 7 earthquake releases 33 times more energy than one of magnitude 6, and roughly 1000 times (33×33) more energy than one of magnitude 5.1994 Northridge was magnitude 6.7. Of course, the site of the earthquake greatly matters, remote area vs populated city for example. Earthquakes of magnitude 6–7 are considered moderate to potentially destructive. 7 to 7.75 are consider major.

The Modified Mercalli scale describes the effect of the earthquake on buildings and people, using Roman numerals. Thus, a level IV is felt by people indoors, a level VII means it would be difficult to stand, furniture would be broken, level VIII means chimneys and factory stacks crumble, etc. etc. . . .

© Springer Nature Switzerland AG 2020
E. Saliklis, *Structures: A Studio Approach*, https://doi.org/10.1007/978-3-030-33153-5_5

The Building's Response

A fundamental fact of seismic loads is this: while the earth shakes violently during an earthquake, the vertical motions are usually not too damaging, especially since architectural engineers readily handle vertical loads in buildings. But the lateral motion is particularly damaging because the earth actually accelerates (i.e. the velocity changes) during an earthquake. Recall that gravity is 32 ft./s^2 this is known as 1 g. It is not uncommon for an earthquake to experience 0.1 g or 0.15 g. Imagine what 0.5 g looks like. A good image is a building cantilevering off of the side of a mountain, and the building feels ½ of its own self weight. That is 0.5 g accelerating horizontally!

The earthquake accelerations can vary from a maximum in one direction, to zero, to a maximum in another direction. So how do seismic loads arrive from ground accelerations? The answer comes from Newton's F = m a. The fact that a building has its own distinctive natural period (the time a structure would take to rock back and forth in free vibration), or its inverse natural frequency is a major factor. If the frequency of the movement of the earth (even though it is random) approaches the natural frequency of the building, response is magnified due to so-called resonance. The most basic such motion such as swaying back and forth, would have the longest period. Other modes of motion exist, but the most basic motion is shown stylized in Fig. 5.1.

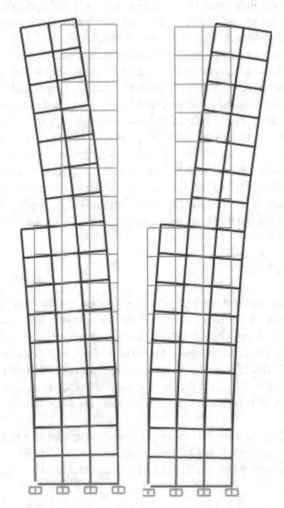

Fig. 5.1 Building swaying in first mode

The Code has come up with a chart that plots the maximum acceleration that a single degree of freedom (SDOF), "lollipop" structure would feel, for varying ground accelerations. Such a structure is simply a lumped mass that is supported by some lateral force resisting system (LFRS), much like a candy lollipop is a ball of sugar supported on a single stick. Such a model allows for a complicated building to be reduced to its first mode response as was shown in Fig. 5.1. Such a building would be subjected to varying ground accelerations, and the first mode response would be captured for a given input acceleration. The worst acceleration that the building feels is noted, and then, either the stiffness or the mass of the building is changed. The same input ground acceleration hits the building and again the worst building acceleration is recorded. A compilation of these worst building accelerations is plotted and that plot is called a Response Spectrum. It is the worst response of the building, over a spectrum of building periods (changing mass or stiffness) for a given seismic event. ASCE7 has such an idealized chart in ASCE Fig. 11.4-1 and it is reproduced here as Fig. 5.2.

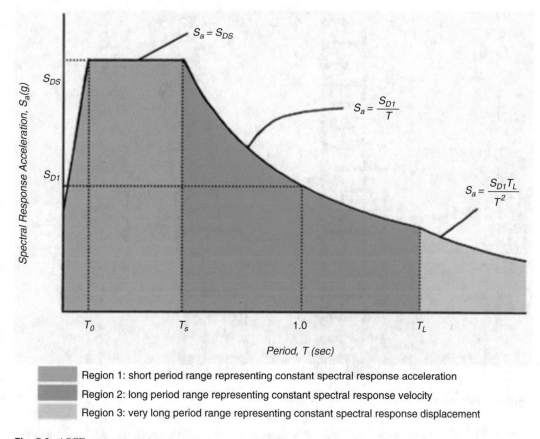

Region 1: short period range representing constant spectral response acceleration
Region 2: long period range representing constant spectral response velocity
Region 3: very long period range representing constant spectral response displacement

Fig. 5.2 ASCE response spectra

For buildings with very, very short natural periods, the motion of the building nearly coincides with the motion of the ground. For buildings with periods of around 0.5 s, (To to Ts in Fig. 5.2), the peak acceleration of the structure can be quite large, in fact it could be several times larger than the acceleration of the ground. As the period of the building increases, say 1–1.5 s, the peak acceleration will be decidedly less than the ground acceleration. The Code translates the range of peak acceleration of a variety of different single degree of freedom (SDOF) buildings to varying ground accelerations into a single curve known as the response spectrum. These accelerations are known as spectral accelerations and they are key to calculating seismic loads.

Certainly the soil conditions have a dramatic effect on the response of a building's acceleration when it is excited by ground motion Site Class A: Hard Rock, B: Rock, C: Very dense soil and soft rock, D: Stiff soil, E: Soft clay soil (Chap. 20 of ASCE). If the soil profile is unknown, it is appropriate to assume Site Class D. In regions of low seismicity, a building on soft soil will accelerate more than if it were on stiff soil. Yet as mentioned before, building with very low natural periods that happen to be on stiff soil or rock would experience large accelerations regardless of the level of seismicity.

Consider a building which is fixed at its base to the ground (Fig. 5.3a). Imagine the earth accelerating horizontally beneath a building to the left (Fig. 5.3b) The building will have a momentary offset at its base, relative to the upper portion of the building. The tendency of the upper portion to remain in its original position is due to inherent inertia.

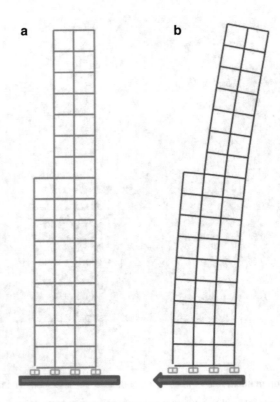

Fig. 5.3 Earthquake loads do not directly strike building

The physical insight derived from the deflected shape of Fig. 5.3b leads to the conclusion that the same distorted or deflected shape could be obtained by applying simulated horizontal forces applied laterally (sideways) to the building. Then, the internal forces and bending moments created by these simulated external forces are equivalent to the forces and bending moments created by the real earthquake ground acceleration. In other words, the simulated external loads are not really applied to a building during an earthquake. They are used to recreate this inertial acceleration effect. There is a strict formula for defining the magnitude of lateral loads to simulate an earthquake. These are shown in a stylized manner in Fig. 5.4. Each level would obtain one load applied to the floor, which is then called a diaphragm, since it receives lateral load.

Fig. 5.4 Equivalent static earthquake loads

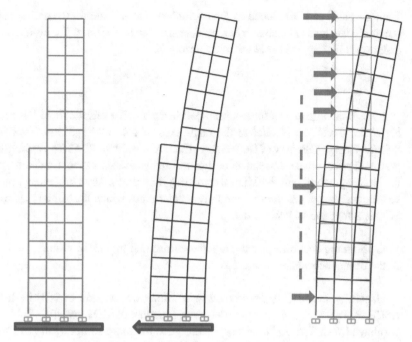

This simulation presents an interesting difference from conventional design ideas. In conventional design, the magnitude of the loads (Dead, Live, Wind etc.) are independent of the stiffness of the building. However, in seismic design, increasing the structure's stiffness increases the induced seismic force, thus the seismic force is dependent on the stiffness of the building! Not only that, but the shape of the building has a profound effect on the resulting seismic forces. What a terrific idea to learn for everyone interested in building design!

And this is the key insight to be used for seismic loads: The simulated lateral forces all add up to a net net base shear V. This V is directly related to mass, but in the Imperial System weight W of the building is used, not mass, but it is a special weight, called the effective seismic weight (ASCE 12.8.1.1).

Then

$$V = C_s \cdot W \tag{5.1}$$

where W is found in ASCE Sect. 12.7.2 and Cs is the seismic response coefficient found in ASCE Sect. 12.8.1.1. From equation ASCE 12.8–2:

$$C_s = \frac{S_{DS}}{\left(\frac{R}{I_e}\right)} \tag{5.2}$$

where S_{DS} is the design spectral response acceleration.

Thus the link between base shear V and the previously defined response spectrum is direct. Notice in Eq. 5–2 the use of the value R which is the response modification factor found in ASCE Table 12.2–1 and the importance factor Ie found in ASCE Sect. 11.5.1.

Think of this idea as follows: $F = mass \cdot acceleration$ is roughly equivalent to $V = S_{DS} \cdot W$ Next, the importance factor is introduced in order to increase V for certain critical structures, thus $V = S_{DS} \cdot W \cdot I$.

Then the value of V is reduced by the coefficient R to account for the ability of the structural system to accommodate loads and absorb energy considerably in excess of the usual allowable stresses, without collapsing. Finally, the basic seismic formula is:

$$V = \frac{S_{DS} \cdot W \cdot I}{R} \tag{5.3}$$

It is worthwhile to pick these terms apart just a bit. The seismic force V is generally evaluated in two horizontal directions parallel to the main axes of the building. One factor in determining V is the Seismic Design Category (SDC). Every structure requires a SDC. The establishment of the SDC has a profound effect on the subsequent design and analysis steps, as it helps determine the structural system, the building height, any building irregularity limitations, any components that must be designed for seismic resistance, and very importantly it determines which horizontal load method you can use. The SDC is a function of two factors:

1. the building occupancy or use and the corresponding risk of failure
2. the design accelerations at the site.

Generally, the SDC is determined twice, first as a function of SDS by IBC Table 1613.3.5(1) or ASCE 7 Table 11.6–1 and second as a function of SD1 by IBC Table 1613.3.5(2) or ASCE 7 Table 11.6–2. The building most be assigned the more severe of the two results. The SDC can be determined based on IBC Table 1613.3.5(1) or ASCE 7 Table 11.6–1 alone if all of the following apply:

- In each direction, the approximate period of the building Ta (a for approximate) is less than $0.8Ts = 0.8SD1/SDS$ This of course means that Ts = SD1/SDS Note that Ts is the period at which the short-period part of the design spectrum transitions into the long-period part of the spectrum. Students sometimes are confused by the units. Ts is not unitless! Units are seconds, even though the ASCE formula implies cancellation of units.
- The seismic response coefficient Cs is determined from: $C_s = \frac{S_{DS}}{\left(\frac{R}{I_e}\right)}$
- The diaphragms are rigid, or for diaphragms that are flexible, the distance between elements of the seismic force resisting system is 40 ft. or less

Notice the thinking here, it is generally unnecessary and wasteful to determine the SDC of a short period building by using long period ground motion.

Now look at what is needed to establish the Risk Category, which is based on the Occupancy Category. In other words, the Risk Category is established from the nature of the occupancy of the building. IBC Table 1604.5 shows how to establish the Risk Category, this is also Table 1.5–1 ASCE 7 and is shown in Fig. 5.5.

TABLE 1604.5
RISK CATEGORY OF BUILDINGS AND OTHER STRUCTURES

RISK CATEGORY	NATURE OF OCCUPANCY
I	Buildings and other structures that represent a low hazard to human life in the event of failure, including but not limited to: • Agricultural facilities. • Certain temporary facilities. • Minor storage facilities.
II	Buildings and other structures except those listed in Risk Categories I, III and IV.

III	Buildings and other structures that represent a substantial hazard to human life in the event of failure, including but not limited to:
	• Buildings and other structures whose primary occupancy is public assembly with an occupant load greater than 300.
	• Buildings and other structures containing Group E occupancies with an occupant load greater than 250.
	• Buildings and other structures containing educational occupancies for students above the 12th grade with an occupant load greater than 500.
	• Group I-2, Condition 1 occupancies with 50 or more care recipients.
	• Group I-2, Condition 2 occupancies not having emergency surgery or emergency treatment facilities.
	• Group I-3 occupancies.
	• Any other occupancy with an occupant load greater than 5,000.[a]
	• Power-generating stations, water treatment facilities for potable water, wastewater treatment facilities and other public utility facilities not included in Risk Category IV.
	• Buildings and other structures not included in Risk Category IV containing quantities of toxic or explosive materials that:
	    Exceed maximum allowable quantities per control area as given in Table 307.1(1) or 307.1(2) or per outdoor control area in accordance with the *International Fire Code*; and
	    Are sufficient to pose a threat to the public if released.[b]

IV	Buildings and other structures designated as essential facilities, including but not limited to:
	• Group I-2, Condition 2 occupancies having emergency surgery or emergency treatment facilities.
	• Ambulatory care facilities having emergency surgery or emergency treatment facilities.
	• Fire, rescue, ambulance and police stations and emergency vehicle garages.
	• Designated earthquake, hurricane or other emergency shelters.
	• Designated emergency preparedness, communications and operations centers and other facilities required for emergency response.
	• Power-generating stations and other public utility facilities required as emergency backup facilities for Risk Category IV structures.
	• Buildings and other structures containing quantities of highly toxic materials that:
	    Exceed maximum allowable quantities per control area as given in Table 307.1(2) or per outdoor control area in accordance with the *International Fire Code*; and
	    Are sufficient to pose a threat to the public if released.[b]
	• Aviation control towers, air traffic control centers and emergency aircraft hangars.
	• Buildings and other structures having critical national defense functions.
	• Water storage facilities and pump structures required to maintain water pressure for fire suppression.

a. For purposes of occupant load calculation, occupancies required by Table 1004.5 to use gross floor area calculations shall be permitted to use net floor areas to determine the total occupant load.

b. Where approved by the building official, the classification of buildings and other structures as Risk Category III or IV based on their quantities of toxic, highly toxic or explosive materials is permitted to be reduced to Risk Category II, provided that it can be demonstrated by a hazard assessment in accordance with Section 1.5.3 of ASCE 7 that a release of the toxic, highly toxic or explosive materials is not sufficient to pose a threat to the public.

Fig. 5.5 ASCE risk categories

Some structures are exempt from earthquake requirements. In particular, the following categories of structures need not be designed to IBC requirements:

• Detached one and two family dwellings where Ss is less than 0.4 g or where the SDC is A, B or C
• Detached one and two family wood frame dwellings not included in the exception above and not more than 2 stories in height. These types of structures must be designed and detailed in accordance with the International Residence Code
• Agricultural storage structures intended only for incidental human occupancy
• Special structures such as bridges, electrical transmission towers, nuclear reactors etc., all of which require special expertise and more rigorous methods

It is important to establish the amount of ground acceleration the building will experience, since as was discussed previously, the acceleration directly relates to forces that the building will experience. A maximum considered earthquake (MCE) has a chance of occurring about once every 2500 years. Contour maps of MCE ground accelerations appear in IBC Figs. 1613.4(1) through 1613.5(14). Ss are mapped accelerations for structures with *s*hort periods (up to 0.2 s) and $S1$ are mapped accelerations for structures with longer periods (*1.0* s). A very useful way of finding out about site specificity of earthquakes is to use the website https://www.seaoc.org/page/seismicdesignmaptool and type in the address of the desired site. The website allows for the consideration of soil conditions.

Typically, engineers ignore the initial slope part of the response spectra curve and just run the line horizontally to the origin, since that portion of the spectrum rarely has an effect on the design of a standard building. That horizontal extension is shown in Fig. 5.6.

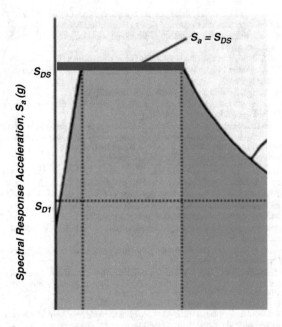

Fig. 5.6 Ignoring first part of response spectra

The very basic formula $V = C_s \cdot W$ stretches back decades in seismic codes. The value of Cs must be established but Cs needs to be calculated for short period range response using S_{DS} and then again for the long period range using S_{D1}. There is a third range but this uses TL and it is a very long period, applicable only to very tall buildings. These will not be studied here.

So Cs can be found at last! From equation ASCE 12.8–2:

$$C_s = \frac{S_{DS}}{\left(\frac{R}{I_e}\right)} \tag{5.4}$$

S_{DS} is found in Sect. 11.4.4 of ASCE7. The steps are a wee bit convoluted. S_{DS} uses short period response but some minimum checks essentially cover the longer period (S_{D1}) case. From ASCE Eq. (11.4–3)

$$S_{DS} = \frac{2}{3} S_{MS} \tag{5.5}$$

and from ASCE Eq. (11.4–1)

$$S_{MS} = F_a S_S \tag{5.6}$$

where Fa is the site coefficient defined in IBC Table 1613.3.3(1) and ASCE7 Table 11.4–1. Fa allows for soft soils to amplify the motion. S_S is the mapped MCE acceleration as determined by Sect. 11.41

which in turn refers to Figs. 22.1, 22.3, 22.5 and 22.6 . . . ouch! It is worth noting that *Ss* is larger near a fault and smaller away from a vault. And if the fault is long, *Ss* is even larger. Values of *Fa* are shown in Fig. 5.7.

STRUCTURAL DESIGN

TABLE 1613.2.3(2)
VALUES OF SITE COEFFICIENT F_v [a]

SITE CLASS	MAPPED RISK TARGETED MAXIMUM CONSIDERED EARTHQUAKE (MCE$_R$) SPECTRAL RESPONSE ACCELERATION PARAMETER AT 1-SECOND PERIOD					
	$S_1 \leq 0.1$	$S_1 = 0.2$	$S_1 = 0.3$	$S_1 = 0.4$	$S_1 = 0.5$	$S_1 \geq 0.6$
A	0.8	0.8	0.8	0.8	0.8	0.8
B	0.8	0.8	0.8	0.8	0.8	0.8
C	1.5	1.5	1.5	1.5	1.5	1.4
D	2.4	2.2[c]	2.0[c]	1.9[c]	1.8[c]	1.7[c]
E	4.2	3.3[c]	2.8[c]	2.4[c]	2.2[c]	2.0[c]
F	Note b	Note b	Note b	Note b	Note b	Note b

a. Use straight-line interpolation for intermediate values of mapped spectral response acceleration at 1-second period, S_1.
b. Values shall be determined in accordance with Section 11.4.8 of ASCE 7.
c. See requirements for site-specific ground motions in Section 11.4.8 of ASCE 7.

TABLE 1613.2.5(1)
SEISMIC DESIGN CATEGORY BASED ON SHORT-PERIOD (0.2 second) RESPONSE ACCELERATION

VALUE OF S_{DS}	RISK CATEGORY		
	I or II	III	IV
$S_{DS} < 0.167g$	A	A	A
$0.167g \leq S_{DS} < 0.33g$	B	B	C
$0.33g \leq S_{DS} < 0.50g$	C	C	D
$0.50g \leq S_{DS}$	D	D	D

TABLE 1613.2.5(2)
SEISMIC DESIGN CATEGORY BASED ON 1-SECOND PERIOD RESPONSE ACCELERATION

VALUE OF S_{D1}	RISK CATEGORY		
	I or II	III	IV
$S_{D1} < 0.067g$	A	A	A
$0.067g \leq S_{D1} < 0.133g$	B	B	C
$0.133g \leq S_{D1} < 0.20g$	C	C	D
$0.20g \leq S_{D1}$	D	D	D

Fig. 5.7 ASCE Fa values

The value for *Cs* need not exceed the following (from ASCE Eq. (12.8–3))

$$C_s = \frac{S_{D1}}{T \cdot \left(\frac{R}{I_e}\right)} \; for \; T < T_L \tag{5.7}$$

Use the SMALLER value of *Cs*, or use *Ts* = SD1/SDS to find that critical transition between short period and the period closer to 1 s, then simply use that appropriate *Cs*.

This of course begs the question, "what is *SD1*"? and the answer is that *SD1* is the design spectral response acceleration for longer periods. So that is the check to ensure *Cs* does not exceed its maximum allowable value. What about the minimum *Cs*? That is in ASCE Eqs. (12.8–5) and (12.8–6).

The value of the fundamental period of vibration of the building may be approximated by:

$$T = C_t \cdot (h_n)^x \tag{5.8}$$

where *Ct* and *x* are given in Table 12.8–2 of ASCE7

The next item to consider is the Response Modification Factor *R*. A structural system is designed to resist dead, live and wind loads and engineers attempt to keep deformations and stresses within acceptably low limits. However, it would be economically prohibitive to use those same limitations when designing for the maximum considered earthquake motion. This is a major source of debate between U.S. Seismic Committees and their Japanese counterparts. How much damage is the owner willing to prevent? In the U.S., the basic philosophy of seismic design is that the structure will be able to accommodate the maximum expected earthquake without collapse. Although the structure is required to ride out the earthquake, inelastic deformation is expected and allowed to occur, as well as structural and nonstructural damage. Thus, the Response Modification Factor *R*, which is determined by the type of lateral load resisting system used, is a measure of the system's ability to accommodate earthquake loads and to absorb energy without collapse. A stiff, brittle structure has a low value of *R*, while a resilient, ductile system has a high value of *R*. The issue of energy absorption is deep and interesting. Start with the premise that the function of the lateral load resisting system is to absorb energy that is induced into the building, and that it does so by moving and deforming, without collapse. Minor damage is expected to be repairable. The ability of structural systems and materials to deform and absorb energy without failure or collapse is called ductility. Thus, materials that can absorb energy through deformations are called ductile, whereas those which are less able to do so are termed non-ductile or brittle.

There are three basic types of lateral load resisting systems: **moment resisting frames** (aka moment frames**), shear walls** and **braced frames**. Generally, shear walls are the stiffest (most rigid), i.e. they deflect the least when subject to a given load. Braced frames are typically less rigid than shear walls, and moment resisting frames are the least rigid and ironically, the most expensive.

A moment resisting frame resists lateral forces primarily through bending action of the columns and of the beams. One such moment resisting frame is shown in Fig. 5.8. Students seeing this for the first time may be surprised that for such a moment frame subjected to lateral load, there might be more energy going into the bending of the horizontal beams, compared to the bending of the vertical columns. In Fig. 5.8, notice that the story drift defines the lateral movement from one story to another, not to some reference like the earth. Also notice that, although all the elements are bending, the original right angles at the joints remain as right angles. This means that the juncture between the beams and the columns must carry bending moment, i.e. they are not hinged. Also notice that the column is typically continuous, i.e. there is no "break" in the column at the first story. Moment resisting frames may be constructed of either structural steel or of reinforced concrete.

Fig. 5.8 Deformation of building due to lateral loads

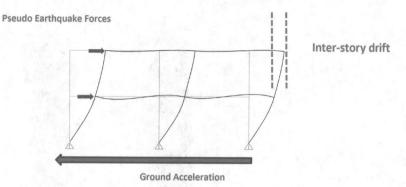

Alas, not all moment resisting frames are alike. There are three recognized categories of moment resisting frames in the Code:

- Special Moment Resisting Frame (SMRF)
- Intermediate Moment Resisting Frame (IMRF)
- Ordinary Moment Resisting Frame (OMRF)

A SMRF is a moment resisting frame made of structural steel or of reinforced concrete that has the ability to absorb a large amount of energy in the inelastic range, that is when the material is stressed above its yield point, and it does so without complete failure and without unacceptable deformation. If a SMRF is concrete, it is almost always cast in place. Note: concrete frames assigned to SDC D and above must be SMRFs. A steel moment frame does not automatically qualify as a SMRF. It must meet specific connection details and it must pass nondestructive testing of the welded connections and only certain grades of steel can be used. The IMRF is a steel or reinforced concrete frame that has less stringent requirements than a SMRF and it generally may only be used in buildings assigned to SDC C and below. An OMRF is a steel or concrete frame that does not meet the special detailing requirements to ensure ductile behavior. For concrete, it is permitted only in buildings assigned to SDC A, B and C.

Shear walls are a commonly used and affordable Lateral Force Resisting System (LFRS). A shear wall resists lateral forces by developing shear in its own plane and by cantilevering from its base. Thus a shear wall is essentially a very deep cantilever beam and as such, it develops axial stress due to bending, as well as shear stress. Of course, the wall tends to lift up at one end, it must be adequately tied down to its foundation. Clearly, the connection of the wall to the foundation must also be sufficient to prevent sliding.

Professors tend to joke that students "lose their mind" when they see a vertical cantilever, and they don't know how to calculate the moment of inertia of such a structure. Is that true? Perhaps the issue boils down to "h" in $bh^3/12$. It is not the height of the wall! This geometry is described in Fig. 5.9.

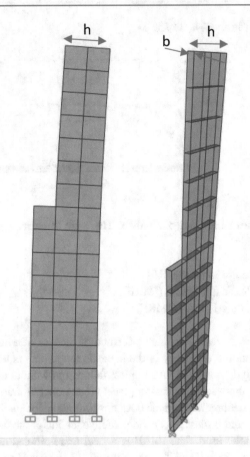

Fig. 5.9 Moment of inertia in shear wall

Understanding the various levels of ductility in different LFRS elements explains some of the ideas behind Table 12.2–1 in ASCE 7 which shows R values for various systems. Note some systems are not limited at all in these applications (NL), whereas some are not permitted to be used at all (NP).

To better understand the forms of braced frames, Figs. 5.10 and 5.11 show a few sketches of common configurations, yet these are all labeled as braced frames. These truly are frames (i.e. with moment carrying joints) and pinned-pinned braces, but sometimes the default assumption is that the entire system is made up of pinned-pinned elements, making these structures 2D trusses for simplicity of hand calculations. Braced frames are most often constructed of structural steel, although reinforced concrete or timber braced frames are possible. There are two types of braced frames, concentric and eccentric. In concentric braced frames (CBF), the centerlines of all intersecting members meet at a point, and consequently the members act primarily as axially loaded elements. The eccentric braced frame (EBF) is a steel braced frame in which at least one end of each brace is eccentric to the center of the beam column joint or the opposing brace. The intent of this design is to make the braced frame more ductile, and thus able to absorb more energy without buckling the braces. Eccentric connectivity is clear on the left of Fig. 5.10 and the left of Fig. 5.11.

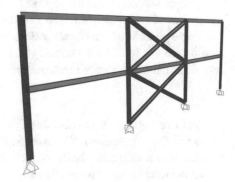
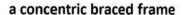

An eccentric braced frame   **a concentric braced frame**

Fig. 5.10 Eccentric braced frames

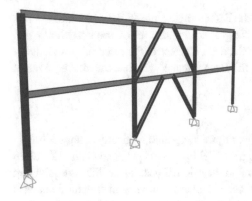

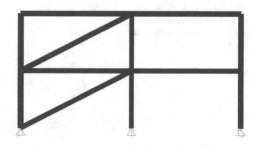
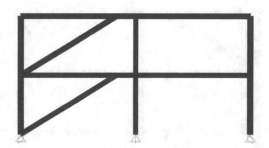

Another eccentric braced frame   **and a concentric braced frame**

Fig. 5.11 More eccentric braced frames

Finally, to determine R, a few more definitions are needed. A Bearing Wall Systems is one in which the shear walls are designed to carry vertical as well as lateral loads. If these elements fail during an earthquake, they won't carry vertical loads and the building might collapse. This explains the low value of R for bearing walls. Height limitations also exist. A Building Frame System is one with an essentially complete frame that provides support for vertical loads. Consequently, lateral loads are resisted by shear walls or by braced frames which carry only minimal vertical loads. A Dual System is one with an essentially complete frame that provides support for vertical loads, then lateral loads are resisted by both moment resisting frames (special and intermediate only) and shear walls or braced frames, in proportion to their relative rigidities. This proportioning of lateral load in accordance with varying stiffness of each LFRS is a very important concept. It will be quantitatively explored in detail. The moment resisting frame acting independently must be able to resist at least 25% of the total required lateral force.

Next, consider the term I_e. The subscript e refers to earthquakes and as mentioned before, the importance factor I_e is found in ASCE Sect. 11.5.1. which is a very brief section because it simply refers to ASCE Table 1.5–2. This table simply links Risk Category to I_e, I = 1.00, II = 1.00, III = 1.25, IV = 1.50

So Cs can finally be found! The last thing needed is the Effective Seismic Weight W which sounds trivial, but of course, it isn't. The Effective Seismic Weight W includes all of the dead loads and applicable portions of other loads as described in ASCE Sect. 12.7.2. Recall the point of these lengthy investigations is to find Cs. The end goal is to establish the Base Shear V which is the total equivalent lateral inertial force from the earthquake.

$$V = C_s \cdot W \tag{5.9}$$

The Code calls for the distribution of the total base shear force V, applied vertically along the height of the building. These inertia forces are found according to ASCE Eqs. (12.8–11) and (12.8–12). Keep in mind that the total of all the lateral Fx forces must equal to the original base shear V. This equivalent static load technique is surprisingly simple and robust. Hand calculations learned in structural analysis courses can be used on such structures of moderate height. The total base shear V is broken up into pieces, according to the height where it will be applied. Each portion of the the base shear V is applied to the diaphragm (the floor), it is not applied to the columns. This distribution of V is a function of the weight at each diaphragm and the height of each diaphragm. If the weights are uniform, then as the height increases, each portion of V increases. This is shown in a stylized manner in Fig. 5.12.

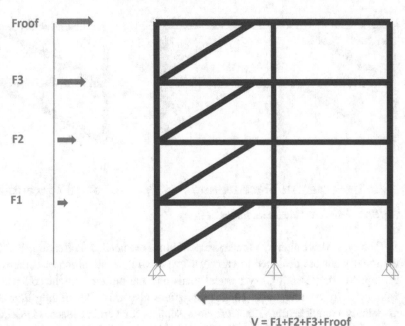

Fig. 5.12 Symbolic depiction of ever increasing equivalent static earthquake loads

The rule for distributing base shear V is found in ASCE Sect. 12.8.3 and is summarized in Fig. 5.13.

Fig. 5.13 ASCE rules for distributing base shear

12.8.3 Vertical Distribution of Seismic Forces

The lateral seismic force (F_x) (kip or kN) induced at any level shall be determined from the following equations:

$$F_x = C_{vx}V \qquad (12.8\text{-}11)$$

and

$$C_{vx} = \frac{w_x h_x^k}{\sum\limits_{i=1}^{n} w_i h_i^k} \qquad (12.8\text{-}12)$$

where

C_{vx} = vertical distribution factor

V = total design lateral force or shear at the base of the structure (kip or kN)

w_i and w_x = the portion of the total effective seismic weight of the structure (W) located or assigned to Level i or x

h_i and h_x = the height (ft or m) from the base to Level i or x

k = an exponent related to the structure period as follows:

for structures having a period of 0.5 s or less, $k = 1$

for structures having a period of 2.5 s or more, $k = 2$

Summary for Earthquake Loads

1. Find the seismic weight
2. Find Cs
3. Find T
4. Find V
5. If there is more than one story, distribute that total V to each diaphragm
6. Calculate how the diaphragm loads travel to the foundation. This includes shear at the base of an LFRS and uplift at the base of an LFRS These last two forces are shown at the bottom of Fig. 5.14.

Fig. 5.14 Base shear and
base T/C couple due to
lateral loads

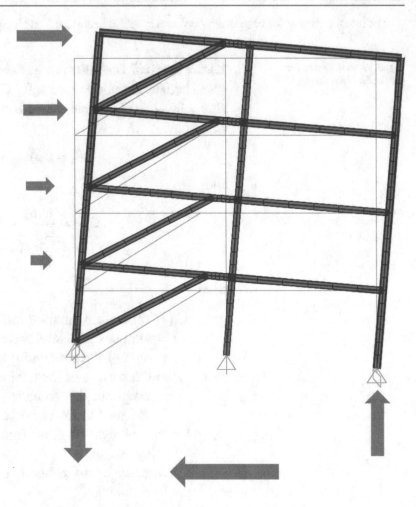

Example Projects to be done in, or outside of studio.

Project 5-1a calculate the effective seismic weight, Cs and base shear (V) for the building

- Find the seismic weight W
- Find SDS and SD1 using the internet tool
- Use Table 12.2–1 to find the Category of the Seismic Force-Resisting System
- Establish Importance I which is linked to Risk
- Find Cs
- Find T
- Cs need not exceed a certain value, check that
- Cs need cannot be below a certain value, check that
- Find V

1a.

Building type	Office
Building dimensions	90' ×80' plan, story heights 14',14',14'
Building weight information	DL Floors & Roof 70 psf – plus partitions at floor Exterior Cladding 20 psf of wall face
Building location	Balboa Zoo, San Diego CA
Soil type	Very dense soil
Lateral system	Special concentric braced frames

Initialization of parametric variables is shown in Fig. 5.15

Fig. 5.15 Initial solution to Project 5-1a

$$\text{Acladding} := (3 \cdot 14\text{ft}) \cdot 90\text{ft} \cdot 2 + (3 \cdot 14\text{ft}) \cdot 80\text{ft} \cdot 2 = 14280\,\text{ft}^2$$

$$\text{Afloor} := (90\text{ft} \cdot 80\text{ft}) \cdot 3 = 21600\,\text{ft}^2$$

$$\text{DL} := 70\frac{\text{lbf}}{\text{ft}^2} \qquad \text{clad} := 20\frac{\text{lbf}}{\text{ft}^2} \qquad \text{hn} := 3 \cdot 14\text{ft}$$

The preliminary steps needed are shown in Fig. 5.16

Fig. 5.16 Next steps to solution for Project 5-1a

Balboa Park SD

Special Concentric Braced Frames

$$W := \text{Acladding} \cdot \text{clad} + \text{Afloor} \cdot \text{DL} = 1797600\,\text{lbf}$$

$$\text{SDS} := 1.127 \qquad\qquad \text{SD1} := 0.478$$

$$V = Cs \cdot W$$

$$Cs = \frac{\text{SDS}}{\left(\dfrac{R}{\text{Ie}}\right)}$$

From Table 12.2-1 we see this is category B2 thus $R := 6$

Importance I comes from Section 11.5.1 fooled ya, really Table 1.5-2 which needs RISK first from Table 1.5-1

$$\text{Risk} = \text{II} \qquad \text{Ie} := 1.0$$

$$Cs := \frac{\text{SDS}}{\left(\dfrac{R}{\text{Ie}}\right)} \qquad Cs = 0.188$$

But Cs NEED NOT exceed the following, assuming we are less than TL which we are

$$\text{Csmax} < \frac{\text{SD1}}{T \cdot \left(\dfrac{R}{\text{Ie}}\right)}$$

The final steps are shown in Fig. 5.17

Fig. 5.17 Final steps to solution for Project 5-1a

We need T   here $hn := \dfrac{hn}{ft} = 42$

$T = Ct \cdot (hn)^x$ from Eqn 12.8-7

$Ct := 0.02$ all other systems in Table 12.8-2

$x := 0.75$ same table

$T := Ct \cdot (hn)^x$ $T = 0.33$ seconds

$Csmax := \dfrac{SD1}{T \cdot \left(\dfrac{R}{Ie}\right)}$ $Csmax = 0.241$

Now check the minimum Cs

$Csmin := 0.044 \cdot SDS \cdot Ie$ $Csmin = 0.05$ $Csabsolutemin := 0.01$

So we are ok $Cs = 0.188$

$V := Cs \cdot W = 337649.2 \, lbf$

Project 5-1b calculate the effective seismic weight, Cs and base shear (V) for the building

Building type	Office
Building dimensions	90' × 80' plan, story heights 14′,14′,14'
Building weight information	DL Floors & Roof 70 psf – Plus partitions at floor
	Exterior cladding 20 psf of wall face
Building location	Balboa Zoo, San Diego CA
Soil type	Very dense soil
Lateral system	Special moment frames

The only difference here is the value of R. Figure 5.18 shows the power of the parametric approach.

Fig. 5.18 Initial solution
to Project 5-1b

$$Cs = \frac{SDS}{\left(\dfrac{R}{Ie}\right)}$$

From Table 12.2-1 we see this is category C1 or C5 thus $R := 8$

Importance I comes from Section 11.5.1 fooled ya, really Table 1.5-2 which
needs RISK first from Table 1.5-1

Risk = II $Ie := 1.0$

$$Cs := \frac{SDS}{\left(\dfrac{R}{Ie}\right)}$$ $Cs = 0.141$

The remaining calculations are simple when they have been previosly programmed. Figure 5.19 concludes this exercise.

Fig. 5.19 Remaining steps
to solution for Project 5-1b

But Cs NEED NOT exceed the following, assuming we are less than TL which we are

$$Csmax < \frac{SD1}{T \cdot \left(\dfrac{R}{Ie}\right)}$$

$$Csmax := \frac{SD1}{T \cdot \left(\dfrac{R}{Ie}\right)}$$ $Csmax = 0.181$

Now check the minimum Cs

$Csmin := 0.044 \cdot SDS \cdot Ie$ $Csmin = 0.05$ $Csabsolutemin := 0.01$

So we are ok $Cs = 0.141$

$V := Cs \cdot W = 253236.9 \, lbf$

Project 5–2 Briefly answer the following questions. Use the ASCE 7 Code to defend your answers.

Question 1 Determine the Risk Category for the following structures:

- A structure containing adult education facilities with 600 person capacity (III)
- Police station (IV)
- A facility that contains extremely volatile and dangerous chemical waste (IV)
- Agriculture building (I)

Question 2 Identify the Seismic Force Resisting Category with a Letter and a Number for the following

Use Table

- Ordinary reinforced masonry shear walls (B 18)
- Prestressed masonry shear walls that act in bearing as well as lateral (A 12)
- Ordinary reinforced concrete moment frames (C 7)

Question 3 Determine the fundamental period of vibration for the following buildings (all story heights are 12 feet):

- 2 story prestressed masonry shear wall building
- Ct = 0.02 x = 0.75

$$\text{2 story prestressed masonry} \qquad hn := 2 \cdot 12$$

$$Ct := 0.02 \qquad x := 0.75$$

$$T := Ct \cdot (hn)^x = 0.217$$

- 4 story steel moment resisting frame building
- Ct = 0.028 x = 0.8

$$\text{4 story prestressed masonry} \qquad hn := 4 \cdot 12$$

$$Ct := 0.028 \qquad x := 0.8$$

$$T := Ct \cdot (hn)^x = 0.62$$

- 2 story glulam timber beam residence

  - Ct = 0.02 x = 0.75

$$\text{2 story glulam} \qquad\qquad hn := 2 \cdot 12$$

$$Ct := 0.02 \qquad x := 0.75$$

$$T := Ct \cdot (hn)^x = 0.217$$

Question 4 Which is more ductile, a building frame system with special reinforced concrete shear walls, or a moment resisting frame system with reinforced concrete special moment resisting frames? In general moment resisting frames are more ductile than shear walls. But let's look closely at ASCE Table 12.2–1 just to make sure. Find the first case, it is B4 it has an R value of 6. Find the second case, it is a SMRF here it is C5, and its R value is 8. The greater the R, the more ductile the system so the moment frame is more ductile than the frame and shear wall system.

Project 5–3 The Lithuanian World Center (that is really the name of it! My distant uncle used to joke "then what is Vilnius? The Universe's World Center?") is in Lemont, IL. It is a multi-use center with classrooms, a basketball court, offices and a Catholic Church. Find the SDC (Seismic Design Category) and also compare the *SDS* and *SD1* from hand calculations versus the internet. Assume very dense soil.

To answer the prompt of Project 5–3 by hand, consider the following steps:

- Establish Risk
- Establish *Ss* and *S1* by hand, use Fig. 22–1 for *Ss* and 22–2 for *S1* these are percent *g*
- Use 11.4–4 to find *SMS* and *SM1* these rely on *Fa* and *Fv*
- Find *Fa* and *Fv* from Table 11.4–1 and Table 11.4–2
- Find the soil modified accelerations
- Find short and 1 s DESIGN accelerations are 2/3 of the soil modified accelerations
- Compare to internet approach

The initialization of variables is shown in Fig. 5.20.

Fig. 5.20 Initial solution to Project 5-3

Lemont World Center Study

Risk Category could be III

The first step is to find Ss and S1 do this via Sect 11.4.1 and in turn Fig 22-2

Site Class C   from 11.4.3   this is the soil

From Fig 22-1 and 22-2 these are percent g

$Ss := 0.11$ $S1 := 0.07$

Use 11.4-4 to find SMS and SM1

$SMS = Fa \cdot Ss$ $SM1 = Fv \cdot S1$

We need Fa and Fv

$Fa := 1.3$ $Fv := 1.5$

Elementary hand calculations are shown in Fig. 5.21

Fig. 5.21 Hand calculation solution to Project 5-3

Thus the soil modified accelerations are

$SMS := Fa \cdot Ss = 0.143$ $SM1 := Fv \cdot S1 = 0.105$

The respective Short and 1sec Design Acclerations are 2/3 of the soil modified acclerations

$SDS := \frac{2}{3} \cdot SMS = 0.095$ $SD1 := \frac{2}{3} \cdot SM1 = 0.07$

Now find the Seismic Design Category from two tables 11.6-1 and 11.6-2 and pick most severe

Short period shows SDC A

1 sec period shows SDC B

Now compare to internet based approach

Comparison to the internet approach is in Fig. 5.22

Fig. 5.22 Internet solution to Project 5-3

Project 5–4 A three story structure has a steel ordinary concentric braced frame at each end of the building. Only this one direction of loading and bracing is studied, namely parallel to gridlines 1 and 2. The building information is shown in Fig. 5.23.

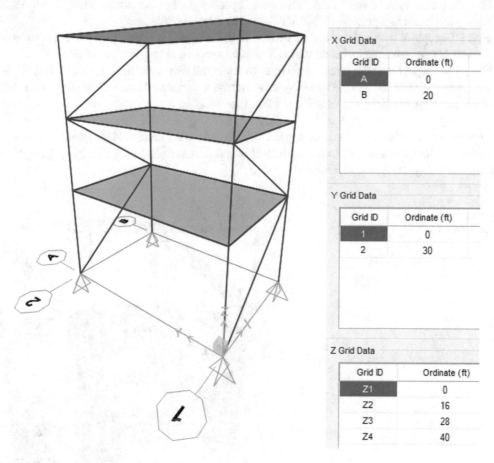

X Grid Data

Grid ID	Ordinate (ft)
A	0
B	20

Y Grid Data

Grid ID	Ordinate (ft)
1	0
2	30

Z Grid Data

Grid ID	Ordinate (ft)
Z1	0
Z2	16
Z3	28
Z4	40

Fig. 5.23 Geometry and spacing Project 5-4

Notice the first story height is larger than the subsequent story height, for architectural openness. Each floor experiences 20 lb./ft^2 LL and the following DL:

- Concrete slab 5/12 foot
- Insulation 2.5 lb./ft^2
- MEP 4 lb./ft^2
- MISC 3 lb./ft^2

Ignore the weight of the braces. Given Importance 1, and SDS $= 0.97$ g and SD1 $= 0.47$ g find what load is applied to each floor. Then find the TC couple at the base of each braced frame ignoring the web members. In SAP2000 run two load patterns, one with DEAD to find Period, one without DEAD to find TC couple.

Be sure to:

- Find Cs based on SDS and find CS based on SD1
- Find the period by hand and by SAP2000 ignoring self weight of the braces (DEFINE > > MATE-RIAL) Cycle through periods till you see flexure in direction of braces!
- Find CT couple at base by hand and by SAP2000. In SAP, look at node with SINGLE brace at base
- Summarize the agreement of hand versus SAP2000 submit a single PDF of all work
- **Start your report** with the summary pictures of the lateral forces and the CT couples in SAP. Start your report with the CT couple forces by hand and the agreement. Then, follow with all the details of the hand calculations used to arrive at those lateral forces.

Typical output of this project is shown in the following figures. Figure 5.24 is a qualitative depiction of axial forces in the braces due to the equivalent lateral forces applied to the building. Blue denotes tension and red symbolizes compression.

Fig. 5.24 Qualitative axial forces Project 5-4

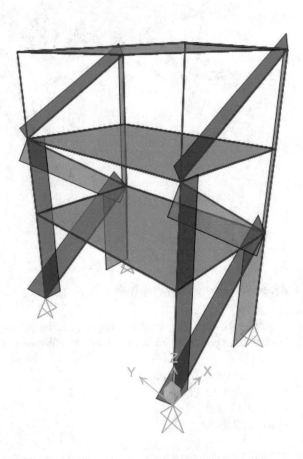

The worst force at a single brace is 60,079 lb. This, and other quantitative axial force information is shown in Fig. 5.25.

Fig. 5.25 Quantitative
axial forces, detail, Project
5-4

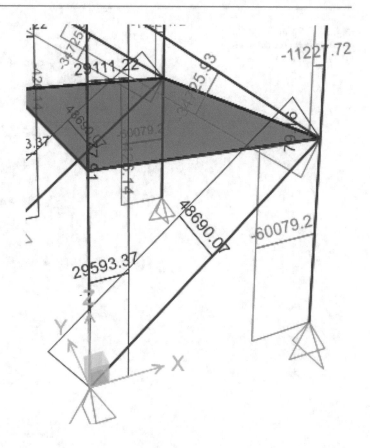

Hand calculations can verify this peak axial force of 60,079 lb.

Give yourself an "atta boy" or atta girl" if these match the SAP2000 values. The verification is in
Fig. 5.26

Fig. 5.26 Verification of
axial forces Project 5-4

$$Mbase := Force1 \cdot Z12 + Force2 \cdot (Z12 + Z23) + Force3 \cdot (Z12 + Z23 + Z34)$$

$$Mbase = 1.202 \times 10^{6}\,\text{ft} \cdot \text{lbf}$$

$$CTcoupleBase := \frac{Mbase}{LAB} = 60078.857\,\text{lbf}$$

The period is found via SAP2000 very easily. Only the floor weights are used for the seismic weight
W. This is shown in Fig. 5.27.

Fig. 5.27 Using SAP2000
to set up Period calculations

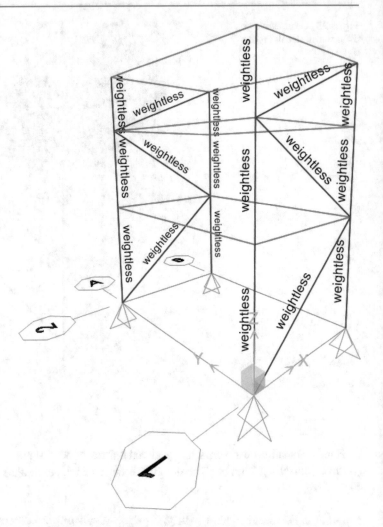

The only tricky part of interpreting the SAP2000 period is that the mode we studied in Fig. 5.3 must be the mode investigated in SAP2000. Depending on many configuration issues, lower modes in SAP2000 may be detected, such as torsional modes. Animating the deformation quickly ensures which mode corresponds to our predicted sway. This happened to be Mode 5 in the model shown in Fig. 5.28. This Mode 5 from SAP2000 had a period of T = 0.2 s, whereas the Hand calculation found 0.294 s. Only 1/10 of a second difference!

Fig. 5.28 SAP2000 output
for first Period

Deformed Shape (MODAL) - Mode 5; T = 0.20628; f = 4.84777

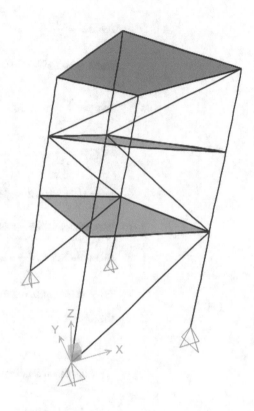

The hand calculations for finding the period are not difficult, nor are they expected to be extremely precise, since the model is a single, simple equation. The Period calculation is shown in Fig. 5.29.

$$\text{LAB} := 20\text{ft} \qquad \text{L12} := 30\text{ft}$$

$$\text{Z12} := 16\text{ft} \qquad \text{Z23} := 12\text{ft} \qquad \text{Z34} := 12\text{ft}$$

$$\text{slabthick} := \frac{5}{12}\text{ft}$$

$$\text{insulation} := 2.5\,\frac{\text{lbf}}{\text{ft}^2}$$

$$\text{MEP} := 4\,\frac{\text{lbf}}{\text{ft}^2} = 0.028\,\text{psi}$$

$$\text{MISC} := 3\,\frac{\text{lbf}}{\text{ft}^2}$$

Ignore LL in seismic weight calculations.

$$\text{floorarea} := \text{L12}\cdot\text{LAB} = 600\,\text{ft}^2$$

$$\text{DL} := \text{slabthick}\cdot150\,\frac{\text{lbf}}{\text{ft}^3} + \text{insulation} + \text{MEP} + \text{MISC} = 72\cdot\frac{\text{lbf}}{\text{ft}^2}$$

$$\text{W1} := \text{floorarea}\cdot\text{DL} = 43200\,\text{lbf}$$

$$\text{W2} := \text{W1}$$

$$\text{W3} := \text{W1}$$

Find the period of the structure, this is not super precise!

$$\text{Ct} := 0.02 \qquad \text{x} := 0.75 \qquad \text{hn} := \frac{\text{Z12} + \text{Z23} + \text{Z34}}{\text{ft}} = 40$$

$$\text{T} := \text{Ct}\cdot\text{hn}^{\text{x}} = 0.318$$

Fig. 5.29   Hand calculations of first Period

Here are the details of the hand calculations to get the three lateral forces. Once these lateral forces are obtained, it is very quick to find the C/T couple at the base of the LFRS. Figure 5.30 sets up the factors needed to distribute the base shear *V* into 3 pieces.

$$R := 3.25$$

$$\text{CsfromSDS} := \frac{SDS}{\dfrac{R}{I}} = 0.298$$

$$\text{CsfromSD1} := \frac{SD1}{T \cdot \dfrac{R}{I}} = 0.59$$

$$Cs := 0.59$$

$$\text{Wtot} := W1 + W2 + W3 = 129600\,\text{lbf}$$

$$V := Cs \cdot \text{Wtot} = 76464\,\text{lbf}$$

$$\text{Voneframe} := \frac{V}{2} = 38232\,\text{lbf}$$

no need for any spreadsheets since there are only three floors

Since the period is less than 0.5seconds use k=1

$$k := 1$$

$$W1h1 := \left(\frac{W1}{\text{lbf}} \cdot \frac{Z12}{\text{ft}}\right)^k \cdot \text{lbf} = 691200\,\text{lbf}$$

$$W2h2 := \left(\frac{W2}{\text{lbf}} \cdot \frac{Z12 + Z23}{\text{ft}}\right)^k \cdot \text{lbf} = 1209600\,\text{lbf}$$

$$W3h3 := \left(\frac{W3}{\text{lbf}} \cdot \frac{Z12 + Z23 + Z34}{\text{ft}}\right)^k \cdot \text{lbf} = 1728000\,\text{lbf}$$

$$\text{denominator} := W1h1 + W2h2 + W3h3 = 3.629 \times 10^6\,\text{lbf}$$

Fig. 5.30 Setting up the vertical distribution of base shear

The three lateral forces all add up to the base shear. Having the three forces allows for a moment equation to be used to find the base C/T couple. This is shown in Fig. 5.31.

Fig. 5.31 Vertically distributed base shear affects base equilibrating forces

$$factor1 := \frac{W1h1}{denominator} = 0.19$$

$$factor2 := \frac{W2h2}{denominator} = 0.333$$

$$factor3 := \frac{W3h3}{denominator} = 0.476$$

$$Force1 := Voneframe \cdot factor1 = 7282 \, lbf$$

$$Force2 := Voneframe \cdot factor2 = 12744 \, lbf$$

$$Force3 := Voneframe \cdot factor3 = 18206 \, lbf$$

$$Mbase := Force1 \cdot Z12 + Force2 \cdot (Z12 + Z23) + Force3 \cdot (Z12 + Z23 + Z34)$$

$$Mbase = 1.202 \times 10^{6} \cdot ft \cdot lbf$$

$$CTcoupleBase := \frac{Mbase}{LAB} = 60078.857 \, lbf$$

Project 5–5 A three story building uses steel, ordinary moment frames to resist seismic loads. The frames are 10 ft. wide, one on each side in the the North/South and East/West direction. The DL is given as 125 lb./ft^2. SDS =0.65 and SD1 = 0.35. Importance = 1. This information is summarized in Fig. 5.32.

Fig. 5.32 Geometry and spacing Project 5-5

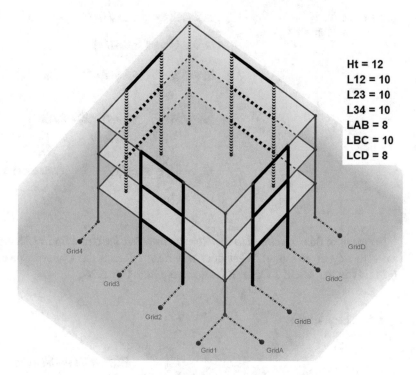

Ht = 12
L12 = 10
L23 = 10
L34 = 10
LAB = 8
LBC = 10
LCD = 8

Grid4

Grid3

Grid2

Grid1

GridA

GridB

GridC

GridD

Tasks:

- Define in simple terms what SDS and SD1 stand for
- Calculate Cs and what Cs need not exceed, use the Code Equations
- Calculate the estimated period T
- Calculate the C/T couple in legs of the frames *at each floor* by hand
- Verify these axial forces at the base of the LFRS via SAP2000
- Draw the qualitative bending moment diagram by hand
- Verify these bending moments via SAP2000

In this problem, we are given *SDS* and *SD1*. These are accelerations to be used for two different scenarios. *SDS* assumes that the period of the building is Short. Spectral Design Short. Yet, we are also given the option of saying that *Cs* need not exceed a certain value. This is based on building periods roughly 1 s. Spectral Design 1 s. The importance for this building under earthquake (*e*) loading is 1, *Ie* = 1. Figure 5.33 shows the beginning of the work.

Fig. 5.33 Initial
calculations for Project 5-5

$$SDS := 0.65 \qquad SD1 := 0.35 \qquad Ie := 1$$

ordinary steel moment frames

$$R := 3.5 \qquad Ct := 0.028 \qquad x := 0.8$$

$$hn := 36 \qquad T := Ct \cdot hn^x = 0.492$$

$$Cs := \frac{SDS}{\left(\dfrac{R}{Ie}\right)} = 0.186 \qquad\qquad Csneednotexceed := \frac{SD1}{T \cdot \left(\dfrac{R}{Ie}\right)} = 0.203$$

The point of this exercise is to stay "on the response spectrum line". Use the SMALLER of the Cs values to stay on this line! What does this mean? Recall the Code Response Spectrum, which is Fig. 11.41 in the ASCE 7 Code. It is shown again in Fig. 5.34.

Fig. 5.34 Deeper study of
response spectrum

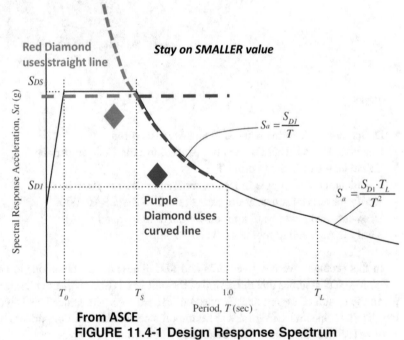

From ASCE
FIGURE 11.4-1 Design Response Spectrum

To stay "on the line" means you have two reasonable choices, the short period part of the curve, or the not-short period part of the curve. Remember that we do not study the long period part. If your structure's period falls on the black diamond in Fig. 5.34, you wouldn't use the short period straight black line, that would not be "on the line", you would use the red curve. If your structure's period coincides with the red diamond in Fig. 5.34, you would not use the red curve, you need to use the black curve to stay "on the line".

Perhaps a simpler way of thinking about this is to simply calculate the Transition Period Ts and figure out which curve governs.

$$T_s = \frac{SD1}{SDS} = 0.54 \ sec \tag{5.10}$$

So use Cs, not Csneednotexceed, since we are left of 0.54.
The base shear is found in the steps shown in Fig. 5.35

Use this smaller value of Cs

Cs := Cs

The base shear V is now readily available.

$DL := 125 \frac{lbf}{ft^2}$ L12 := 10ft   L23 := 10ft   L34 := 10ft

LAB := 8ft   LBC := 10ft   LCD := 8ft

AreaFloor := (LAB + LBC + LCD)·(L12 + L23 + L34) = 780·ft^2

WtFloor := DL·AreaFloor = 97500·lbf

Wtot := 3·WtFloor = 292500·lbf

V := Cs·Wtot = 54321.43·lbf

Fig. 5.35 Solution of base shear for Project 5-5

The next step is to distribute portions of this total base shear to each diaphragm. The formula is easy to use, notice the load increases as you go up. These are reproduced from ASCE 7 in Fig. 5.36.

Fig. 5.36 Distribution of
base shear per Code

12.8.3 Vertical Distribution of Seismic Forces

The lateral seismic force (F_x) (kip or kN) induced at any level shall be determined from the following equations:

$$F_x = C_{vx}V \qquad (12.8\text{-}11)$$

and

$$C_{vx} = \frac{w_x h_x^k}{\displaystyle\sum_{i=1}^{n} w_i h_i^k} \qquad (12.8\text{-}12)$$

where

C_{vx} = vertical distribution factor

V = total design lateral force or shear at the base of the structure (kip or kN)

w_i and w_x = the portion of the total effective seismic weight of the structure (W) located or assigned to Level i or x

h_i and h_x = the height (ft or m) from the base to Level i or x

k = an exponent related to the structure period as follows:

for structures having a period of 0.5 s or less, $k = 1$

for structures having a period of 2.5 s or more, $k = 2$

No need for any spreadsheet since there are only three stories. Since the period is less than 0.5 s we can use k = 1. The factors used to divide up the total shear are shown in Fig. 5.37

Fig. 5.37 Calculating factors to distribute base shear

$Ht := 12\text{ft}$

$\text{W1H1} := \text{WtFloor} \cdot \text{Ht}$ $\qquad$ $\text{W2H2} := \text{WtFloor} \cdot 2 \cdot \text{Ht}$ $\qquad$ $\text{W3H3} := \text{WtFloor} \cdot 3 \cdot \text{Ht}$

$\text{denom} := \text{W1H1} + \text{W2H2} + \text{W3H3}$

$\text{CV1} := \dfrac{\text{W1H1}}{\text{denom}} = 0.17$ $\qquad$ $\text{CV2} := \dfrac{\text{W2H2}}{\text{denom}} = 0.33$ $\qquad$ $\text{CV3} := \dfrac{\text{W3H3}}{\text{denom}} = 0.5$

Use the defined distribution factors to distribute the load vertically. This simple calculation is shown in Fig. 5.38

$$V3 := CV3 \cdot V = 27160.71 \cdot lbf$$

$$V2 := CV2 \cdot V = 18107.14 \cdot lbf$$

$$V1 := CV1 \cdot V = 9053.57 \cdot lbf$$

Fig. 5.38 Using distribution factors to find lateral diaphragm loads

The C/T couples are quickly found now Use ½ of these shears as there are two frames resisting lateral load. Frame width is L23 ft. The hand calculations are shown in Fig. 5.39.

Fig. 5.39 Using lateral loads to find base equilibrating forces

$$V3 := CV3 \cdot V = 27160.71 \cdot lbf$$

$$V2 := CV2 \cdot V = 18107.14 \cdot lbf$$

$$V1 := CV1 \cdot V = 9053.57 \cdot lbf$$

$$CTlevel3 := \frac{\dfrac{V3}{2} \cdot Ht}{L23} = 16296 \cdot lbf$$

$$CTlevel2 := \frac{\dfrac{V3}{2} \cdot (Ht + Ht) + \dfrac{V2}{2} \cdot Ht}{L23} = 43457 \cdot lbf$$

$$CTlevel1 := \frac{\dfrac{V3}{2} \cdot (Ht + Ht + Ht) + \dfrac{V2}{2} \cdot (Ht + Ht) + \dfrac{V1}{2} \cdot Ht}{L23} = 76050 \cdot lbf$$

The C/T couple at the base of the LFRS, which was just found to be 76,050 lb. is verified with an elementary SAP2000 model. The output of the model is shown in Fig. 5.40.

Fig. 5.40 SAP2000 verification of base axial forces

Trusses

6

Trusses are extremely efficient structures because they carry a variety of loads primarily through axially loaded members. Although trusses are studied in introductory courses, the linkages of truss mechanics via Virtual Work in Analysis Classes, and via finite element tools in Computer Labs, will increase a technical understanding, as well as broaden one's visual literacy. Insights from structural analysis courses allow trusses to be distilled down to simple chords acting in tension and compression to carry bending, and simple forces in the web members to carry shear. Calculating the internal moment in a bent truss allows for changing that internal moment like that of a beam, to a C/T couple of forces at some distance (arm) from each other. This C/T couple insight will be helpful in explaining some structural behavior.

The basic premise of considering a planar truss, as being comprised of pin-ended axially loaded members that are loaded only at the nodes (joints), must be adhered to. Such a physics model does not allow for loading anywhere on a truss other than at the nodes which connect elements. Yet the overall truss bends like a beam, with tension on one side and compression on the other side. The bent truss is still composed of straight members, they do not distort in bending.

Such trusses can be small scale vertical planar systems supporting a floor such as shown in Fig. 6.1, or horizontal diaphragm systems as shown in Fig. 6.2.

Fig. 6.1 Vertical truss supporting horizontal floor

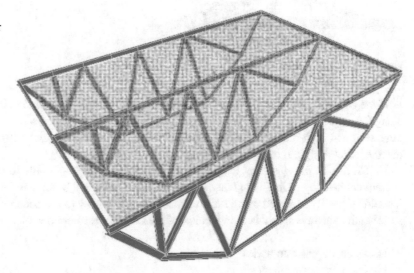

© Springer Nature Switzerland AG 2020
E. Saliklis, *Structures: A Studio Approach*, https://doi.org/10.1007/978-3-030-33153-5_6

Fig. 6.2 Horizontal truss
as rigid diaphragm

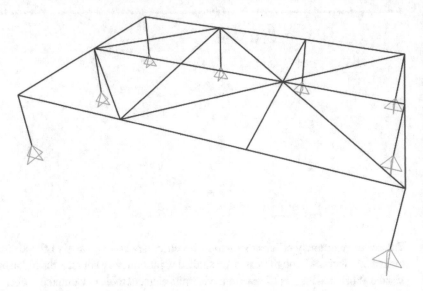

Begin by looking at elementary trusses which are formed by triangular units, which can be further subdivided into more triangular units. Figure 6.3a shows a classical truss with a point load at a node, Fig. 6.3b shows the deformed shape greatly exaggerated and Fig. 6.3c shows a qualitative axial force diagram induced by the external load.

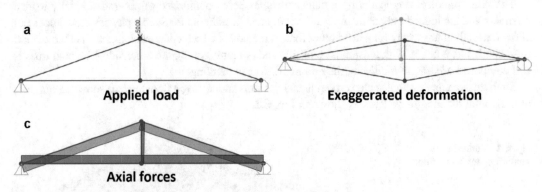

Fig. 6.3 Qualitative behavior of simple truss

If the classical rules are followed, then such a truss can only be loaded at its nodes, not anywhere along the member length. If the classical rules are followed further, then each member is assumed to be pin-ended and each individual member cannot bend. This rule is applicable to timber design, and to much of steel design. It is not applicable to reinforced concrete design.

In real life, as the spans get longer and longer, it becomes impossible to neglect self-weight and subsequent bending of individual members. This is not what will be studied for hand calculations, but it could be readily analyzed using finite element tools. It is easy to see bending in the distorted shape, such bending always signifies the violation of the classical assumptions of:

- Loads only applied to nodes
- Pin-ended truss members

Such complicated, but more realistic bending of a non-classical truss is shown in Fig. 6.4. Avoid such bending when creating a classical truss model ! If such bending exists, it certainly means that the moment carrying capabilities at the ends of truss members were not released or that self-weight of the truss is not being ignored.

Fig. 6.4 Self weight clearly not ignored

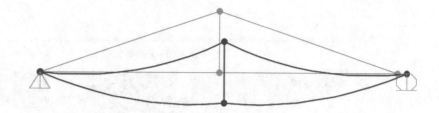

Recall that trusses can span great distances and that they can carry large gravity loads. They are efficient because each individual pin-ended member acts as a two-force member with axial loads only, no bending of any individual member. But what about the truss as a whole, how does it deform? Clearly on a macro scale, a horizontal truss looks a lot like traditional beams do! Consider the following truss made up of five "panels" that are 12 length units wide and 6 length units deep. The truss is subjected to a floor load (force/length^2) on its top chord, which is changed to a uniformly applied load (force/length). But that load is then changed to a series of point loads (force) applied to the top chord nodes. Now look how the entire truss deflects under these uniform point loads. This is shown in Fig. 6.5.

Fig. 6.5 Overall deformation shows truss acts as a beam

If a free body cut was made at some point along the length of this structure, an acceptable model of behavior is to change what would have been the internal bending moment of the analogous beam, into a T/C couple composed of only chord forces at the very top and bottom of the truss.

Figure 6.6 shows one such analogous beam model. This one simply uses the distributed load (force/length) along the top. In Fig. 6.6, the distributed load w is 0.1 force/length, and the span length L is 60 units of length. The peak moment is immediately found from the elementary equation

$$M_{peak} = \frac{w \cdot L^2}{8} \tag{6.1}$$

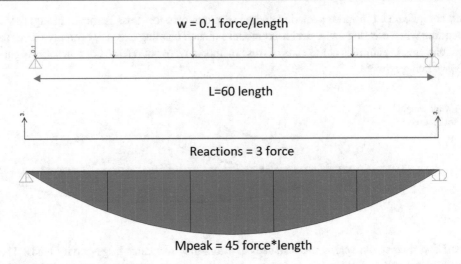

Mpeak = 45 force*length

Fig. 6.6 Simple beam under uniformly distributed load

The internal bending moment varies along the length, from the truss-as-beam analogy. At any particular cut, the moment can be found. That moment is reduced down to a T/C couple, which acts at the arm length of the depth of the truss. If the arm length is constant as in Fig. 6.7, the T/C couple forces in the chords vary directly in response to the variation of the moment.

Fig. 6.7 C T Couple changes in response to magnitude of internal moment

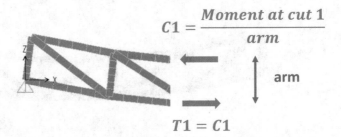

$$C1 = \frac{Moment\ at\ cut\ 1}{arm}$$

arm

$$T1 = C1$$

Moment at cut 1, beam analogy

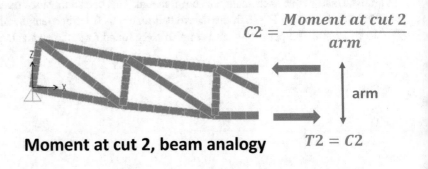

$$C2 = \frac{Moment\ at\ cut\ 2}{arm}$$

arm

Moment at cut 2, beam analogy $T2 = C2$

Notice that these moments, at any cut, arose from a uniformly distributed force on an analogous beam. It is perfectly appropriate to model uniformly distributed loads as discrete point loads, if the top nodes are somewhat closely spaced to each other along the top chord of the truss. That is the preferred approach. Compare the bending moment diagram of Fig. 6.6, which arose from a uniform load assumption, with the bending moment diagram of Fig. 6.8, which arose from the more precise placement of loads only at the nodes. The difference is not worth the effort of calculating the piecewise linear bending moment diagram of Fig. 6.8, compared to the parabolic diagram of Fig. 6.6.

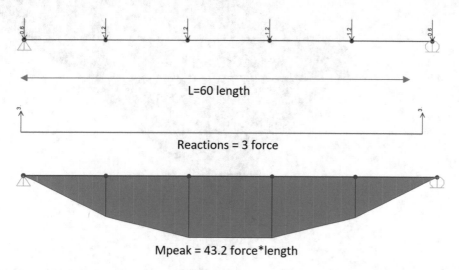

Fig. 6.8 Simple beam under closely spaced point loads

A more interesting observation is the "halving" of load at the extreme left and right edges of the top chord in Fig. 6.8. This arises from the typical situation of a truss being uniformly loaded, thus an end node on such a truss only has half the tributary width of an interior node.

But the most useful and important observation is this: How did the uniform load arrive to the top chord of the truss in Fig. 6.6? The answer to this question lies in Fig. 6.9. Here, a series of parallel trusses span some large distance. They all support a floor stylized as a red slab over the trusses in Fig. 6.9. The applied floor load will be denoted as force/length^2. The challenge is to convert that force/length^2 load into a uniform force/length load, to be applied to the analogous truss-beam. Then, if desired, that uniform load of force/length can be broken up (discretized) into individual loads (force) on each top node, with a halving of that load at the extreme first and last nodes. But that discretization is not necessary for closely spaced loads such as those shown in Fig. 6.9.

Fig. 6.9 Closely spaced
top nodes implies
uniformly distributed load
model for moment

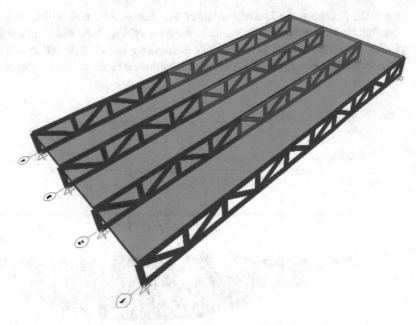

The tributary width of load flowing into any interior truss is half the distance to the adjacent truss on the left and half the distance to the adjacent truss on the right. This width is of course a length, thus the load on an interior truss is:

$$uniform\ load = floorload \cdot tribwidth \qquad\qquad (6.2a)$$

$$\frac{force}{length} = \frac{force}{length^2} \cdot length \qquad\qquad (6.2b)$$

Notice the use of gridlines in Figs. 6.9 and 6.10. All project submittals must have clear, unambiguous gridlines. Letters in one direction, numbers in the other direction. Figure 6.10 shows the intuitive approach of deciding how much load flows to an individual truss. Such load flow is based on the so-called "tributary width" associated with an individual structural load bearing member.

Fig. 6.10 Tributary width
of an individual truss
acting as beam

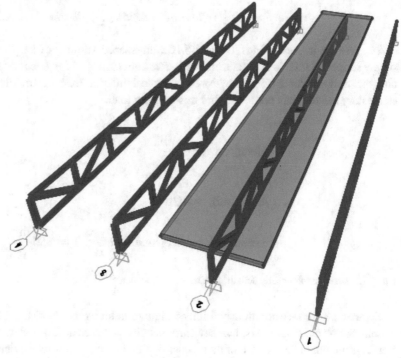

Why is the area associated with an element ½ the span to the next member? The Müller-Breslau Method provides an answer. Figure 6.11 shows the Müller-Breslau construction for the supporting end reaction of an interior truss holding up a floor load. The reaction end is lifted up 1 unit of height, and two areas ensue over the fold. Area 1 has a ½ drop and a ½ drop, as does Area 2.

Fig. 6.11 Müller-Breslau
approach to establishing
tributary width

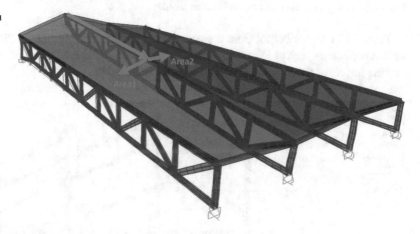

Thus the load flowing into one end of the "beam truss" is:

$$Reaction_{end} \cdot 1 = floorload \cdot Area1 \cdot \frac{1}{2} \cdot \frac{1}{2} + floorload \cdot Area2 \cdot \frac{1}{2} \cdot \frac{1}{2} \tag{6.3a}$$

$$Reaction_{end} \cdot 1 = floorload \cdot Area1 \cdot \frac{1}{4} + floorload \cdot Area2 \cdot \frac{1}{4} \qquad (6.3b)$$

Suppose the trusses of Figs. 6.9 and 6.10 were spaced 16 units of length apart. The span of each simply supported truss is 100 units of length. The floor load is 35 force/length^2. All units are consistent, find the uniform load (force/length) on a typical interior truss-beam. Figure 6.12 describes the elementary calculations needed to find this uniform load.

$$floorload \ := 35 \frac{lbf}{ft^2}$$

$$tribwidth := 16ft$$

$$beamload := floorload \cdot tribwidth = 560 \frac{lbf}{ft}$$

Fig. 6.12 Changing floor force/length^2 to truss-beam force/length

Figure 6.13 shows a common real-life configuration for a truss. Notice in Fig. 6.13, that the top and bottom chords are continuous, i.e. they carry bending moment across nodes. Notice also that the web members are pin-ended at each of their ends, signified by the gap between lines. This means that there is no bending at all in any web member, but both the top and bottom chords do indeed bend, because of their continuity across nodes.

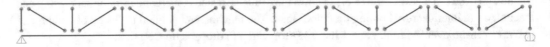

Fig. 6.13 Truss with continuous top and bottom chords

Figure 6.14 shows output from a small, parametric modeling program. "Parametric" means that certain parameters such as long span length, spacing between trusses, depth of truss, etc. can be varied, and the impact of these variations is documented and studied.

Fig. 6.14 Initial steps of parametric model of truss-beam

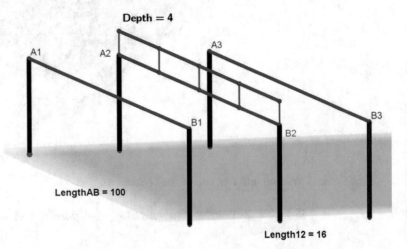

When creating such output, it is very important to have clean, informative graphics. Gridlines need not be shown as this is not a construction drawing, but some identifying labels tied to gridlines are necessary, here the end nodes are labeled as intersections of two gridlines. Line weight is chosen based on aesthetics and is consistent. A minimal uncluttered palette is desired. When programming for such a parametric study, the variable names must be meaningful (tribwidth, Length12, LengthAB, floorload, etc.), this will help with debugging.

After the parametric mini-study which focuses exclusively on the T/C couple forces, the next step is to size the area of the top or the bottom chord based on an allowable stress. Here the units get muddy in the Imperial System. To use feet and lb (pounds force=lbf) consistently, allowable stress can be described using feet or inches. Here are typical allowable stresses for structural steel.

$$\sigma_{allowable} = 5184000 \; \frac{lbf}{ft^2} \tag{6.4a}$$

$$\sigma_{allowable} = \frac{Force}{area} \tag{6.4b}$$

In the U.S. inches are always used to specify cross sectional areas. For pounds and inches use:

$$\sigma_{allowable} = 36000 \; \frac{lbf}{in^2} \tag{6.5}$$

Continuing the previous example which established the uniform load as 560 lb/ft along a typical interior truss which had a 4ft depth, the sizing of the top and bottom chords is quickly found via the following calculations shown in Fig. 6.15. Notice that the programming language used in Fig. 6.15 distinguishes between pounds of force (lbf) and pounds of mass (lb). Here, lbf must be used.

Fig. 6.15 Design of top or bottom chord of truss-beam by hand

$$span := 100ft$$

$$Mworst := \frac{beamload \cdot span^2}{8} = 700000 \, lbf \cdot ft$$

$$depth := 4ft$$

$$TorC := \frac{Mworst}{depth} = 175000 \, lbf$$

$$\sigma allowable := 36000 \frac{lbf}{in^2} \qquad \sigma allowable = 5184000 \frac{lbf}{ft^2}$$

$$Area := \frac{TorC}{\sigma allowable} = 4.861 \cdot in^2 \qquad Area = 0.034 ft^2$$

Calculations which repeat frequently, which is the essence of a parametric study, must be done in computational environment, not by hand. Digital calculations also allow for very neat and clean summaries of the steps taken, thus a checker can more clearly understand the logic used. Figure 6.16

shows the output of the previous truss, but this time a few sliders were manipulated to change some variables and the final area required is immediately found. Always thoroughly debug such a program before beginning any parametric variations.

Fig. 6.16 Design of top or bottom chord of truss-beam by parametric model

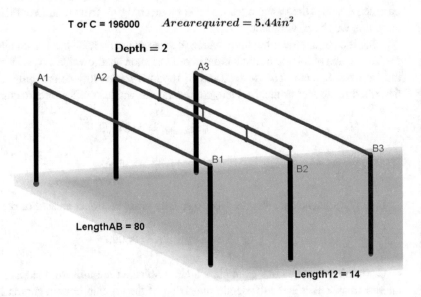

$$\text{T or C} = 196000 \qquad Area required = 5.44 in^2$$

$$Depth = 2$$

A1   A2   A3

B1   B2   B3

LengthAB = 80

Length12 = 14

Next, consider a truss with a large central span, and two smaller but mirror-symmetric overhangs. The tops of the trusses support a roof, which is not shown, but the roof experiences a load of force/length^2. The trusses are evenly spaced. This is shown in Fig. 6.17.

Fig. 6.17 Series of truss-beams with two overhangs

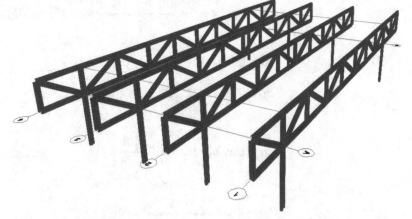

Suppose a typical interior truss experienced 500 lb/ft along its top chord. The central span is 60 units of length, and each overhang is 12 units of length. Recall that this uniform load arises from the floor load (force/length^2) multiplied by the tributary width (length) which here is the truss spacing. The columns at gridlines A and B have been replaced with a pin and roller respectively, to create an analogous beam model. This analogous beam model is shown in Fig. 6.18.

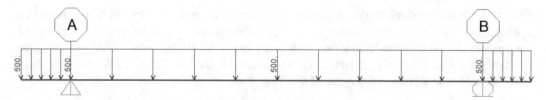

Fig. 6.18 Analogous beam model of a typical interior truss

Figure 6.19 shows the exaggerated deflected shape of the analogous beam, Fig. 6.20 shows the qualitative bending moment diagram, and Fig. 6.21 shows the quantitative bending moment diagram. Notice how small the negative, cantilever moment is compared to the positive moment at the center of the span. Drawing bending moments to some comfortable scale, especially in a computational drawing environment, immediately allows the designer to see the differences in moments, even for a qualitative drawing such as Fig. 6.20.

Fig. 6.19 Exaggerated deflected shape of the analogous beam

Fig. 6.20 Qualitative bending moment diagram of the analogous beam

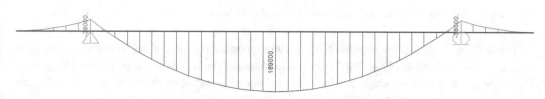

Fig. 6.21 Quantitative bending moment diagram of the analogous beam

The physics of the bending may seem complicated, but one can superpose two very simple situations because of the mirror-symmetry of this problem. Such problems can always be broken up into a "free moment" case and a "restraining moment" case, and the final state is the superposition of these two cases. If the load is uniformly distributed, then the free moment case will always be a parabola, with a peak value shown in Eq. (6.6).

$$M_{peak\ parabola\ free\ moment} = \frac{w \cdot L^2}{8} \tag{6.6}$$

Where w is the vertical distributed load (force/length) and L is the horizontal span of the segment being studied. Here L is the column to column distance between gridlines A and B. The only "trick" is that the parabola can be distorted, with either a starting point or an ending point matching up with the restraining moment. Figure 6.22 shows the loading scenario of the "free" case and Fig. 6.23 shows the corresponding bending moment diagram of the "free" case.

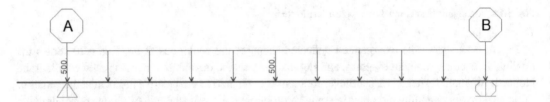

Fig. 6.22 Loading scenario of the "free" case of the analogous beam

Fig. 6.23 Qualitative bending moment diagram of the "free" case of the analogous beam

Figure 6.24 shows the loading scenario of the "restraining" case.

Fig. 6.24 Loading scenario of the "restraing" case of the analogous beam

The next step is subtle, but important. The bending moment diagram of the "restrained" case must be drawn at the same scale as the "free" case to make sense of the superposition of the two. Figure 6.25 shows the bending moment diagram of the "restraining moment" case. Notice the central, unloaded portion has constant moment. There is no load in the central span in this second case.

Fig. 6.25 Qualitative bending moment diagram of the "restraining" case of the analogous beam

The parabolic free case is lifted to match the restraining moments over each column, this is done at the left end, gridline A and the right end, gridline B, because there is a restraining moment at each of these points. The superposition is shown in Fig. 6.26.

Fig. 6.26 Superposition of both bending moment diagrams

Figure 6.27 shows the completed analysis.

A parametric study can now be performed, wherein the central span can be changed, the length of the cantilever overhangs can be changed, the depth of the truss can be changed and the load can be changed. Mirror symmetry of the truss about the central point will allow for the superposition previously described. This superposition will readily find the peak bending moment, then the corresponding T/C couple can be found and the top and bottom chords can be sized based on an allowable stress. Of course, the allowable stress itself can be varied for even more investigations. Two configurations are shown in the following figures. Figure 6.27 shows a central span of 60ft with 12ft overhangs, a 500 lb/ft load and a 3ft truss depth. In Fig. 6.27, the red Xs are used to "spline" a circumcircular arc, which is not technically correct as the shape is a parabola, but circular arcs are extremely easy to program and the resulting answers are nearly perfectly accurate, the peak moment matches that of Fig. 6.20 with the peak moment at the center being 189,000 lbf ft

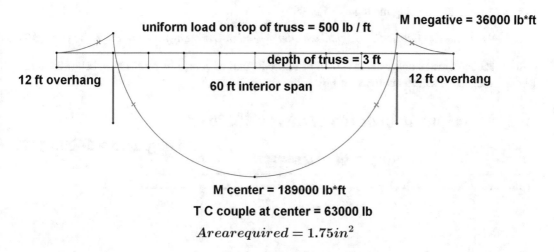

Fig. 6.27 Completed analysis on truss-beam with two overhangs

With the push of a few sliders, a different geometry can be studied and solved for immediately once the parametric program is debugged. Figure 6.28 shows a central span of 30ft with 12ft overhangs, but with a 2ft deep truss and a 500lb/ft load along the top chord.

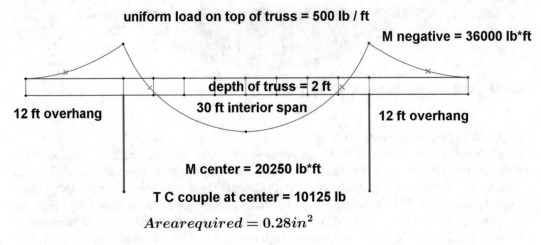

Fig. 6.28 New analysis output from parametric program

The moment at the center of the truss of Fig. 6.27 can be found in an almost ridiculously easy manner using the Müller-Breslau method. To find that moment, the analogous beam is cracked at the point of investigation and a unit rotation is induced at that crack. The centroid of the load on the right side is easily found, and the loft of that load is readily established through similar triangles. The central moment then is found using the logic of Eq. (6.7).

$$M_{central} = floorload \cdot (RightCentralSpan + RightOverhang) \cdot OverallLoft \cdot 2 \qquad (6.7)$$

This Müller-Breslau approach is shown in Fig. 6.29 and of course, it precisely matches the more laborious algebraic statics approach of Fig. 6.27.

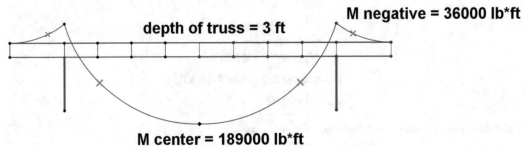

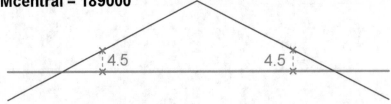

Fig. 6.29 Müller-Breslau approach to finding moment at center of interior span

Project 6–1 A typical prompt given to students follows.

The first part of this project is to be done by hand, preferably on a serviette (table napkin), but of course using graph paper is acceptable. This is a parametric study of a large span truss. The parameters to vary are:

- The left overhang
- The right overhang
- The central span
- The depth of the truss
- The uniform load (lb/ft) which is on the entire top chord this symbolizes DL + LL
- The point load at the left overhang tip
- The point load at the right overhang tip

The goal of the study is to draw the bending moment diagram and to find the T C couple force that acts at the center of the truss. Note this may not necessarily be the worst case, but that is the goal today.

Draw a few scenarios by hand. But as always, first debug your thinking. Photograph or scan your serviette sketches. Figure numbers and figure titles are below each figure. Use clear, unambiguous text with units describing the worst moment at each overhang, and the final central moment, and the final T C couple.

Extra credit if you can detect the peak moment and the peak TC couple which may not be in the center!

Figure 6.30 shows output from a parametric program created with GeoGebra.

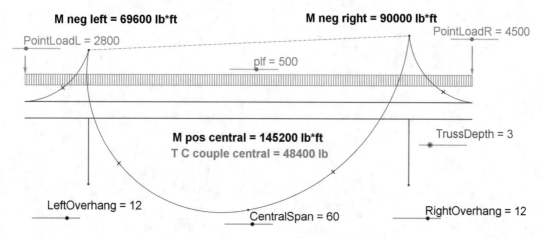

Fig. 6.30 Typical output for Project 6–1

A finite element program such as SAP2000 could be used to check values such as the moments at the start of the overhang. Figure 6.31 shows typical SAP2000 output. The top of Fig. 6.31 shows the truss as a beam. The middle figure is the displaced shape and the lower figure is the bending moment diagram which exactly matches the Müller-Breslau approach.

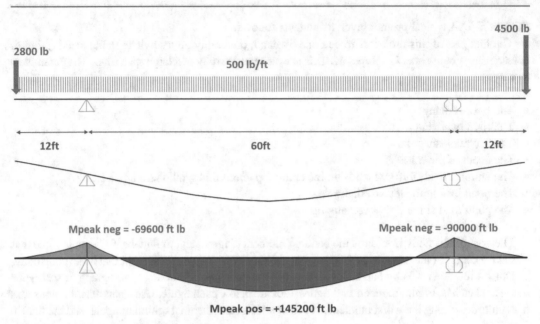

Fig. 6.31 Finite element verification of Project 6–1 output

Flexible Diaphragms 7

Consider now the situation of reinforced concrete, or reinforced masonry shear walls supporting a plywood diaphragm. Relatively speaking, one system (the wall) is much, much stiffer than the other system (the roof diaphragm). In this situation, we look at the structural system as a beam (the roof diaphragm) being flexed between immovable (rigid) wall supports. This structural system is classified as a flexible diaphragm. In reality, nothing is really perfectly rigid, but engineers assume one of the following two situations to be true. Either:

- The diaphragm is rigid and the supporting walls are flexible
  Or
- The diaphragm is flexible and the supporting walls are rigid.

Figure 7.1 shows a flexible diaphragm structural system. Notice that even though the interior shear walls are not as long as the exterior shear walls, they are all the same stiffness because all of the walls are assumed rigid, whereas the diaphragm is assumed to be flexible.

Fig. 7.1 Four rigid shear walls laterally support one flexible diaphragm

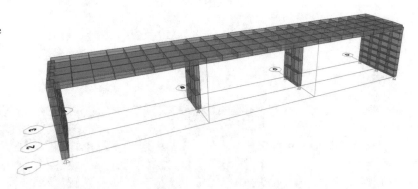

To emphasize that the diaphragm is indeed flexing like a beam, consider the exaggerated deformation of the structure subjected to a lateral load applied to the diaphragm. This is shown in Fig. 7.2. The shear walls are picking up load from the diaphragm. In this situation the diaphragm is flexible and the shear walls are rigid, so the shear walls do not move, they are like pinned supports restraining a bent beam.

© Springer Nature Switzerland AG 2020
E. Saliklis, *Structures: A Studio Approach*, https://doi.org/10.1007/978-3-030-33153-5_7

Fig. 7.2 Diaphragm bends in its own plane between rigid supports

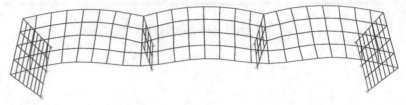

This behavior is fundamentally different than the rigid roof problem, which is studied in great depth in the next chapter. If the roof is rigid, the shear walls are by default flexible, each with its own rigidity. There, the roof would move as a unit, if the problem is symmetric, and the load picked up by each shear wall in the symmetric problem would be a function solely of the walls relative stiffness. This is shown in Fig. 7.3, with load being distributed in proportion to the wall stiffness.

Fig. 7.3 Two scenarios of four flexible shear walls supporting one rigid diaphragm

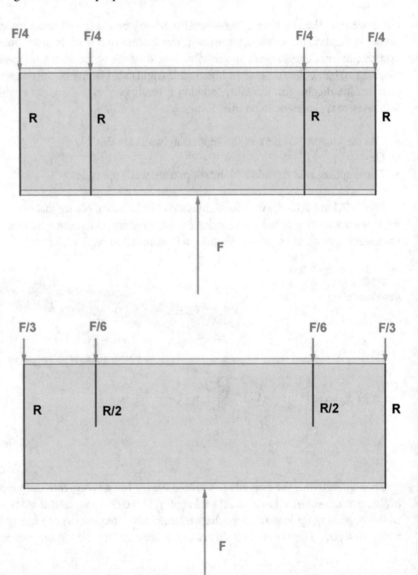

But in the flexible roof problems studied in this chapter, all the walls are rigid, regardless of their length. This is shown in Fig. 7.4. Notice in Fig. 7.4 the very important element known as a "collector" or "drag strut" shown with a dashed line. It anchors the flexible diaphragm in place, and it connects under the diaphragm, rigidly attaching to the shear wall. The inner walls take twice what the outer walls carry.

Fig. 7.4 Four rigid shear walls, two collectors, three segments of flexible diaphragm

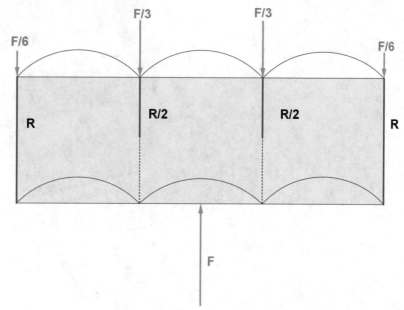

Figure 7.3 showed two rigid diaphragms in plan view. Compare the rigid diaphragm statics of Fig. 7.3 to the flexible diaphragm statics of Fig. 7.4. A rigid diaphragm, supported by four equally stiff shear walls, is shown in the upper part of Fig. 7.3. Since the diaphragm moves as a unit rigidly, each shear wall deforms the same amount and thus the force induced in each wall ($F = k \Delta$) is identical. In the rigid diaphragm on the bottom of Fig. 7.3, the inner walls are half as stiff as the outer walls, thus, under rigid roof deformation, the inner walls are induced to experience half of the force that the outer walls experience. However, in Fig. 7.4 we have the new situation of a flexible diaphragm supported by (relatively) rigid shear walls. Thinking of tributary width, it becomes apparent that the inner walls have twice the tributary width responsibility that the outer walls have. Thus, it makes sense that they pick up twice the load as the outer walls, even if they have a different rigidity than the outer walls do, as shown here. If shear walls are really stiff compared to the diaphragm, they act as pin supports restraining the lateral load transferred by the flexible diaphragm.

Note there is something quite important about Fig. 7.4, that is the fact that the inner shear walls do not extend fully under the diaphragm. In order for the diaphragm to fully transfer its force to these inner shear walls, there must be an element that connects the diaphragm to the shear wall. That element is called a drag strut, or sometimes it is called a collector. It transfers load from the diaphragm to the shear wall and is shown as a dashed line in the plan view of Fig. 7.4. No drag struts is needed between the diaphragm and the outer walls in Fig. 7.4 because the walls extend fully under the diaphragm all the way from north to south. The force in the drag strut linearly increases from zero at the south end to some peak compressive force when it hits the shear wall. If the load was reversed and was moving south, the drag strut forces linearly decrease from some peak tensile load at the point where it connects to the shear wall, and reduces to zero when it terminates at the south end. Why is that? To answer this question consider Fig. 7.5. In this diagram there are many "rigid" shear walls supporting a flexible diaphragm roof (not shown) and many drag struts.

Fig. 7.5 3D assembly of shear walls and collectors

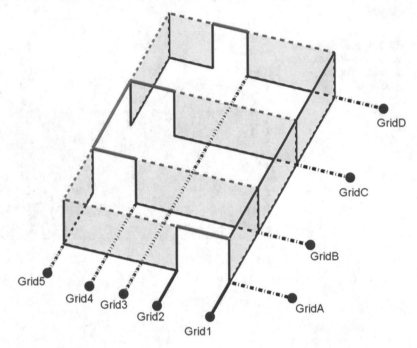

For lateral loading parallel to the numbered grids (1, 2, 3 etc.)., the walls aligned with the numbered grids 1 and 5 do all the resisting. They do so by acting as beams cantilevering up from the ground, and they bend in their strong direction. The walls aligned with lettered grids do nothing to resist load parallel to the numbered grids because they would flop over for such loads, they are not acting in the strong direction, so we assume they do not resist lateral loads at all.

Conversely, for loads in the lettered grid direction we have a different situation. For loading in the direction of A, B, C etc., the shear walls aligned with grids A, B, C and D are engaged and resist loads. The walls perpendicular to the load do nothing to resist loads. Thus, for loads parallel to the lettered grids, the wall along grid1 and grid5 do nothing at all, we assume they do not exist.

Next, imagine a lateral load moving parallel to the numbered grids in Fig. 7.5. If this were wind, the windward load might push, with a tributary area, onto the face in line with grid A. That wind load would be transferred to the diaphragm (not shown) and walls aligned with grid 1 and grid 5 resist. But then the diaphragm (not shown) continues over the gap in the walls along grid 5. The diaphragm is connected to the drag struts all along grid 5. For this loading, the strut between grids B and C along line5 experiences load. But when that load finally flows along the dashed red line in Fig. 7.5 all the way to the face on grid D, the perpendicular wall cannot resist anything. Thus, the axial force in the drag strut must be zero when it hits the perpendicular wall. Similarly, the axial force at point A5 must be zero, for this load along the numbered lines, as the wall on A cannot resist perpendicular load.

Referring back to Fig. 7.2, it is apparent that one side of the flexible diaphragm experiences tension, and the other side experiences compression. Tensile stresses and strains develop one side and compressive stresses and strains develop on the other side because the diaphragm acts as a beam in bending, and there is some internal moment M. These stresses are exactly the same ones studied in earlier classes, namely axial stresses induced by bending.

The moment M induces a Tension/Compression couple, and the tension force developed must balance the compression force. It is impossible to balance stresses, one only can balance forces. The tensile force is some distance, or 'arm' away from the compressive force.

Figure 7.6 shows a bent element, like our flexible diaphragm, being flexed by a bending moment. This induces compression on one side of the neutral axis and tension on the opposite side of the neutral axis. The net compression is symbolized by C acting not at the centroid of the compressed area, but rather at the extreme edge. Similarly, the net compression is symbolized by T, which is placed not at the centroid of the tensile triangle, but rather, at the extreme edge. This is shown in Fig. 7.6, starting with a moment, then with a stress profile and finally with a $T\,C$ couple.

Fig. 7.6 C T Couple is at extreme edges of stress block, not at centroids

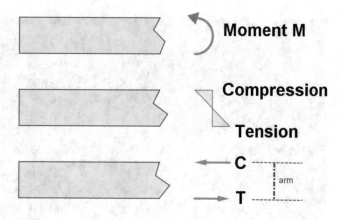

The equivalency of the $C\,T$ couple and the moment M can be demonstrated a number of ways.

$$M = C \cdot arm = T \cdot arm = C \cdot \frac{arm}{2} + T \cdot \frac{arm}{2} \tag{7.1}$$

Now consider the same bending as applied to the flexible diaphragm. The forces developed in the flexible diaphragm are known as "chord forces", because there is an actual need for some structural element, analogous to the top and bottom chord of a truss, to be responsible for these C and T forces. This is shown in Fig. 7.7.

Fig. 7.7 C T Couple is at extreme edges of flexible diaphragm

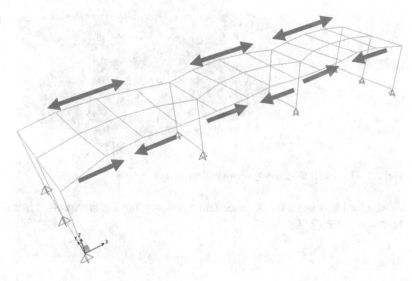

It is not practical to place these chords at some interior spot, so it is traditional to simply assign chords to the extreme edges of the flexible diaphragm. The distance between these two chords is the arm described in Fig. 7.6. The arm distance is the length of diaphragm, or think of it as the total depth of the flexed element. This explains why the C and the T force is not placed at the centroid of the triangular stress distribution profile of Fig. 7.6.

A summary of the load flow concepts associated with flexible diaphragms follows. These wonderful drawings were created by Professor Yasushi Ishida. Figure 7.8 re-emphasizes that drag struts are needed only where the shear wall is no longer in contact with the diaphragm. This occurs for architectural reasons.

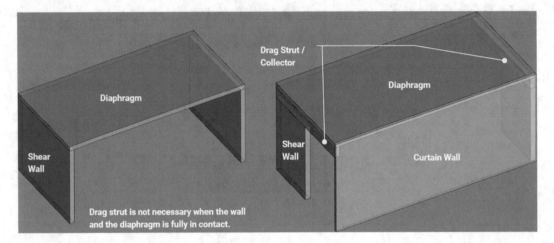

Fig. 7.8 Overview of lateral load flow, initial geometry

It cannot be emphasized enough that shear walls only work as in-plane elements. When they are loaded out-of-plane they are assumed to have zero stiffness. This is highlighted in Fig. 7.9.

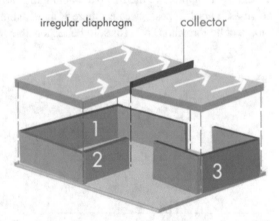

Fig. 7.9 Collector is needed only where wall terminates

The load flow begins with the curtain wall accepting a load expressed in units of force/length^2. This is shown in Fig. 7.10.

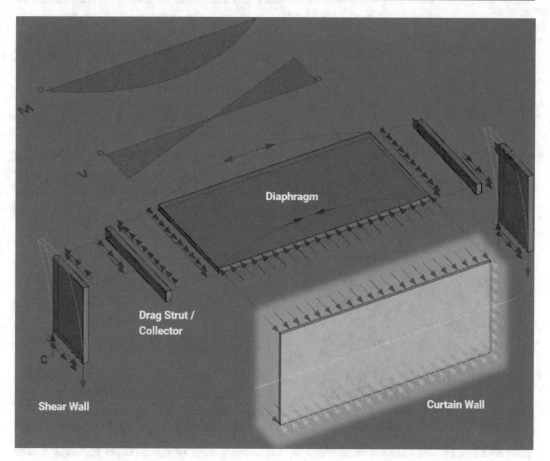

Fig. 7.10 Wind hits façade

Typically, the curtain wall is supported at its base, so only ½ of the total wind load on a one story building flows into the diaphragm as a force/length. This is shown in Fig. 7.11.

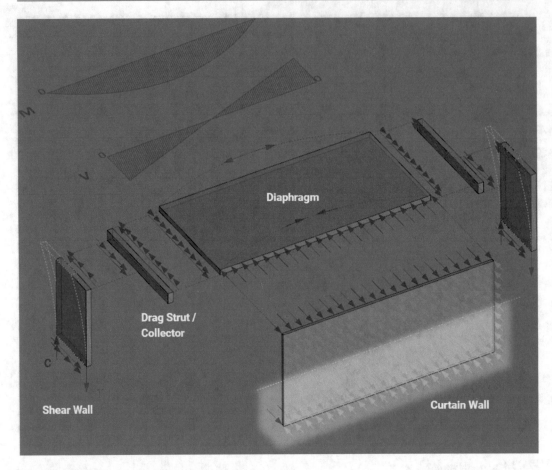

Fig. 7.11 Half of the wind load flows to foundation

The other tributary width of load flows into the diaphragm. Force/length^2 becomes force/length when multiplied by the tributary width. This is shown in Fig. 7.12.

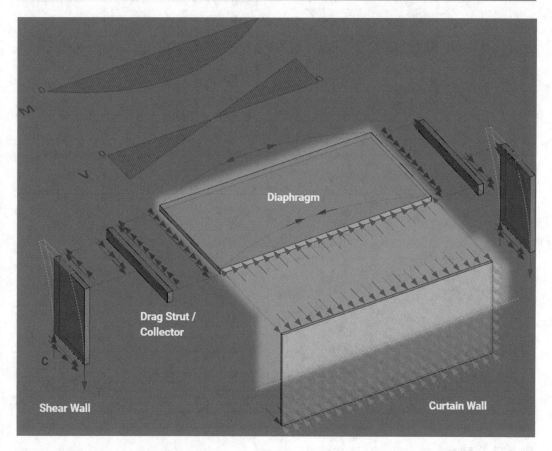

Fig. 7.12 Half of the wind load flows to the diaphragm

The diaphragm flexes as a simply supported beam would. The shear walls act as pinned or hinged supports and compression develops on the curtain wall side and tension on the opposite side. This is shown in Fig. 7.13. Notice the classic shear and bending moment diagrams.

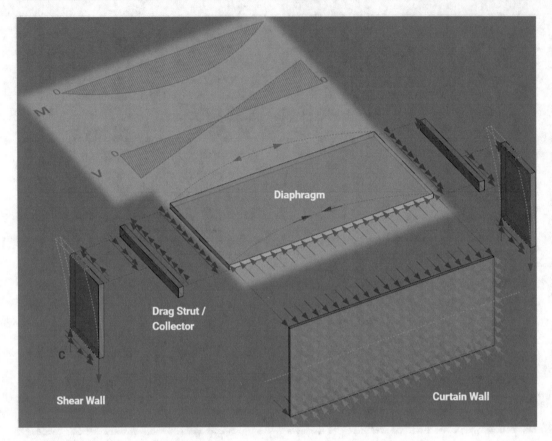

Fig. 7.13 Diaphragm flexes as a simply supported beam would

Even though we assume the connection is pinned, the diaphragm is fully connected to drag struts, (or collectors) along its left and right edges. Notice the direction of shear flow on the underside of the diaphragm. It resists movement on the underside, and the units of this shear flow are force/length. That force/length shear flow is passed to the top of the drag strut/collector, equal in magnitude but opposite in direction. This is shown in Fig. 7.14.

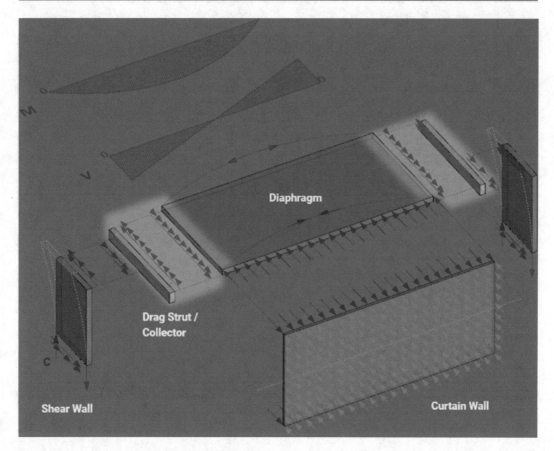

Fig. 7.14 Two different shear flows act on drag strut/collector

This shear flow is once again transferred from the bottom of the drag strut to the top of the shear wall. Notice that if the length of contact is smaller on the underside of the drag strut, the magnitude of shear flow there must be greater than it is on the top of the drag strut for equilibrium to occur. This is shown in Fig. 7.15.

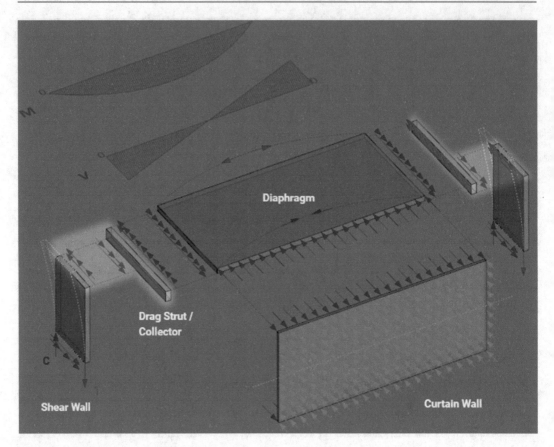

Fig. 7.15 Shear flow from bottom of drag strut/collector goes to wall

The shear flow on the underside of the drag strut is transferred to the top of the shear wall. Notice that these are equal in magnitude, but opposite in direction as expected. This is shown in Fig. 7.15. The shear wall is a vertical cantilever and it also experiences base shear. The cantilever bending it experiences is about its strong axis. That is why the assumption of out-of-plane stiffness is zero, the moment of inertia out-of-plane is very small, while the in-plane moment of inertia is very large. The base moment is resolved into a $T\,C$ couple, with a tensile force at one corner and a compressive force at the other corner. Notice this occurs on the left side and on the right side of the flexible diaphragm as shown in Fig. 7.16.

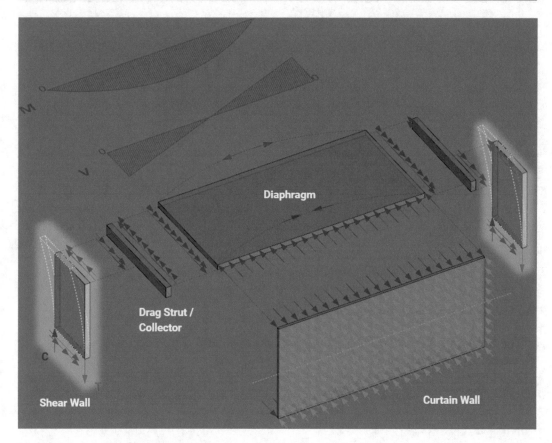

Fig. 7.16 Shear flow on wall induced C T Couple at base

A numerical example of load flow would trace the lateral load, through the flexible diaphragm to the $C\ T$ couple at the extreme edges of the bent diaphragm. Then the rigid walls would support the diaphragm, but these walls would be bent as cantilevers in their own planes, so they too would develop a different $C\ T$ couple at their base. Consider the flexible diaphragm shown in Fig. 7.17. It is subjected to a uniformly distributed load of 250 lb./ft. hitting the diaphragm in the pure Y direction. The bay widths in the x direction are 28 ft. each and the length of the building in the y direction is 20 ft. The height of the building is 12 ft. Calculate:

- The maximum bending moment in the plane of the flexible diaphragm.
- The maximum chord force (tensile or compressive) developed in the chords
- The chord force (tensile or compressive) one third span from wall
- The base shear in each of the rigid shear wall
- The T/C at the base of the inner shear wall

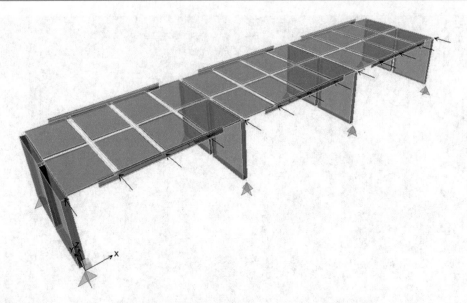

Fig. 7.17 Geometry of flexible diaphragm numerical example

The calculations for each of the above steps are not complicated at all. They are shown in Fig. 7.18 via elementary algebraic statics.

Fig. 7.18 Algebraic steps to find C T Couple at base

$$wind := 250\,\frac{lbf}{ft} \qquad span := 28ft \qquad armroof := 20ft$$

$$Mmid := \frac{wind \cdot span^2}{8} = 24500 \cdot lbf \cdot ft \qquad TCroofmid := \frac{Mmid}{armroof} = 1225 \cdot lbf$$

$$Vleft := \frac{wind \cdot span}{2} = 3500 \cdot lbf$$

$$onethird := \frac{span}{3} = 9.333 \cdot ft \qquad Monethird := Vleft \cdot onethird - wind \cdot \frac{onethird^2}{2}$$

$$Monethird = 21777.8 \cdot lbf \cdot ft$$

$$TCroofonethird := \frac{Monethird}{armroof} = 1088.9 \cdot lbf$$

$$Vright := \frac{wind \cdot span}{2} = 3500 \cdot lbf \qquad Vtopinnerwall := Vright + Vleft = 7000 \cdot lbf$$

$$widthwall := 20ft \qquad heightwall := 12ft$$

$$Momentbaseinnerwall := Vtopinnerwall \cdot heightwall = 84000 \cdot lbf \cdot ft$$

$$TCinnerwall := \frac{Momentbaseinnerwall}{widthwall} = 4200 \cdot lbf$$

The previous concepts can be quantified in another example. The structure shown in Fig. 7.19 has gridlines A, B, C and D (in the x direction) which are 20 ft. apart, and gridlines 1, 2, 3 (in the y direction) which are also spaced at 20 ft. The height of the building is 12 ft. (z direction). The wind is blowing on cladding (not shown) on gridline 1. The cladding is secured to the ground and to the diaphragm, thus one half of the wind load will go directly to the ground and the other half will go to the flexible roof diaphragm. The wind load is 25 lb/ft^2 (psf).

Fig. 7.19 Another numerical example problem

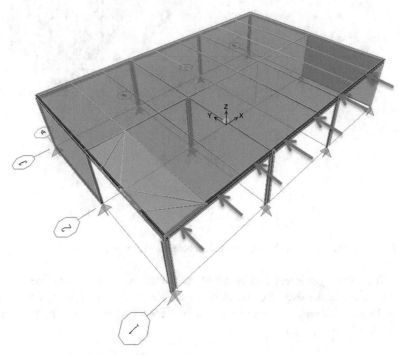

Along gridlines A and D in Fig. 7.19, there is a shear wall extending for 20 ft. Then for the remaining 20 ft. there is a drag strut (also known as a collector) which takes the job of transferring the diaphragm loads from the diaphragm where there is no wall, to the shear walls. Drag struts cannot transfer forces to columns. For example, neither the column at 1A, nor the column at 3D can carry lateral load.

We can calculate the shear per linear foot (force/length) that the diaphragm experiences, these are the shear arrows shown in Fig. 7.20.

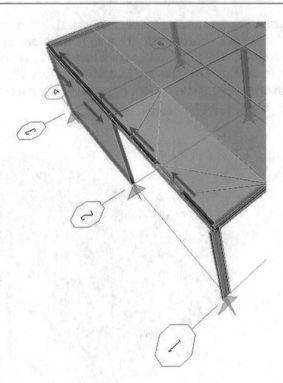

Fig. 7.20 Shear flow graphically depicted on collector and on wall

This shear per linear foot is known as the unit shear and is represented by v_{roof}. It is simply the magnitude of the shear force divided by the length (not the span!) of the diaphragm. Students often make a mistake here. Be sure to use the roof length when calculating v_{roof}, and to use the wall length when calculating v_{wall}.

$$v_{roof} = V_{max}/length\ of\ diaphragm \qquad (7.2)$$

Next, consider the drag strut which lies in between the diaphragm and the shear wall. This drag strut is responsible for transferring all of the load at either the left or the right of the flexible diaphragm (i.e. *Vmax* of Fig. 7.13) to the shear wall. The top of the drag strut feels v_{roof} which is coming from the bottom of the diaphragm. The bottom of the drag strut feels v_{wall}, which is what gets transferred to the shear wall. Of course, if the wall extends under the entire bottom of the drag strut, then these two unit shears are identical. But many times, for architectural reasons the shear wall is not as long as the drag strut.

Certainly the forces must equilibrate, i.e. the total force which is V_{max} must be picked up by the bottom of the drag strut. Otherwise horizontal equilibrium will not be maintained. Thus, v_{wall}, i.e. the unit shear transferred from the bottom of part of the drag strut to the top of the shear wall is found as:

$$v_{wall} = V_{max}/length\ of\ shear\ wall \qquad (7.3)$$

Lastly, this v_{wall} is transferred to the top of the shear wall. This induces an equilibrating shear at the base of the shear wall and strong axis bending in the shear wall, symbolized by the *C/T* couple.

The next step which puts it all together conceptually may be the hardest for students to grasp, but keep in mind that it is a re-telling of what was just described. Figure 7.21 shows the two shear flows. The shear arrows on the bottom of the diaphragm (and on the top of the drag strut) are v_{roof}. The shear arrows on the bottom of the drag strut (and on the top of the shear wall) are v_{wall}. A free body diagram (FBD) of the drag strut helps plot the variation of axial force in the strut induced by v_{roof} above the drag strut and v_{wall} below the drag strut. This is shown in Fig. 7.21.

Fig. 7.21 Shear flow on
collector and associated
axial force diagram

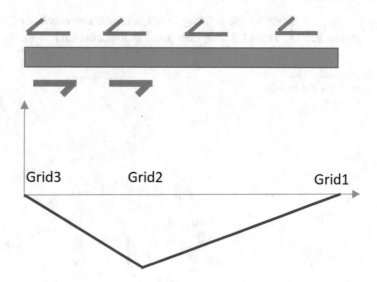

The net axial force must be the superposition of v_{roof} and v_{wall}. In Fig. 7.21, the strut starts out with zero axial force on gridline 1. Next, it picks up compressive force/length, increasing in compression, peaking as it hits the wall on gridline 2. Then, between gridlines 2 and 3, it starts to change and gradually be equilibrated by v_{wall}. Since the total axial force must equilibrate (that induced by v_{roof} and that induced by v_{wall}), the axial force goes back to zero at gridline 3.

Project 7–1 A few quantitative questions follow. Figure 7.22 shows a structural grid in plan view. The wind hits the southern curtain wall and blows to the north with some force/length^2. This has been changed to 400 lb/ft through the tributary width of the curtain wall. The wind also hits the western curtain wall and blows to the east with some force/length^2. This has been changed to 160 lb/ft through its tributary width. The solid black lines in Fig. 7.22 are shear walls and the thin red lines are drag struts.

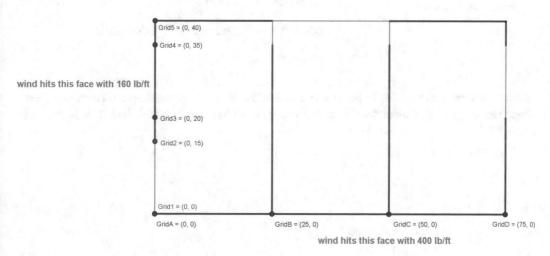

Fig. 7.22 Project 7–1 geometry and loading plan view

Figure 7.23 shows the building in 3D, with the shear walls and the red collector elements (drag struts). Each wind load is truly distributed as lb/ft., but each is shown as a dashed single arrown.

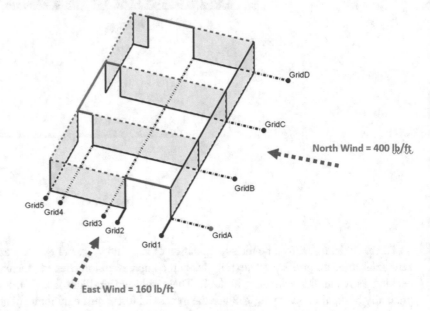

Fig. 7.23 Project 7–1 geometry and loading 3D view

Question 1 For eastern wind load only, what is the magnitude of the axial force in the strut along Grid5 at the B5 intersection?

To do this in a parametric environment, first set up all the variables. This is always done at the beginning, not in the body, of any programming script. An example is shown in Fig. 7.24

Fig. 7.24 Initial steps for parametric model answers to Project 7–1

$$\text{weast} := 160 \, \frac{\text{lbf}}{\text{ft}} \qquad \text{wnorth} := 400 \, \frac{\text{lbf}}{\text{ft}}$$

$$L12 := 15\text{ft} \qquad L23 := 5\text{ft} \qquad L34 := 15\text{ft} \qquad L45 := 5\text{ft}$$

$$LAB := 25\text{ft} \qquad LBC := 25\text{ft} \qquad LCD := 25\text{ft}$$

Find the total load pushing to the east and divide that by 2 to get the simple shear at either end. These elementary calculations use the parametric variables previously established. This is shown in Fig. 7.25.

$$\text{EastTotForce} := \text{weast} \cdot (\text{L12} + \text{L23} + \text{L34} + \text{L45})$$

$$\text{EastTotForce} = 6400 \cdot \text{lbf}$$

$$\text{VEastWind} := \frac{\text{EastTotForce}}{2} = 3200 \text{ lbf}$$

Fig. 7.25 Elementary calculations to find shear force at each end

Find the shear flow on the underside of the roof. Notice the use of lower case v for shear flow, and upper case V for shear force.

$$vEastRoof = \frac{VEastWind}{LAB + LBC + LCD} = 42.6 \; lb/ft \tag{7.4}$$

The next step is slightly tricky! To find the shear flow in the wall, consider all of the walls supporting the roof on that underside of the diaphragm, not one or the other wall, but all of the walls. In this problem that would be the length between gridlines A and B plus the length between gridlines C and D.

$$vEastwall = \frac{VEastWind}{LAB + LCD} = 64 \; lb/ft \tag{7.5}$$

Figure 7.26 explains the equilibrium of the drag strut. When considering *vroof*, it is easiest to go from the top downward. The wind hits the diaphragm and pushes it to the right. Under the diaphragm, *vroof* resists sliding to the left. That drags the top of the strut to the right. When considering *vwall*, it is easiest to go from the bottom up. The base shear resists the wind to the left. *Vwall* hits the top of the wall going to the right, thus *vwall* drags the strut to the left. It is easiest to assume the normal force in the drag strut is tensile.

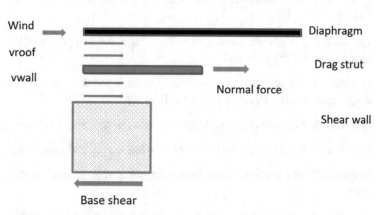

Fig. 7.26 Exploded view of lateral load path

$$\text{Normal} + \text{vEastroof} \cdot \text{LAB} - \text{vEastwall} \cdot \text{LAB} = 0$$

$$\text{Normal} := \text{vEastwall} \cdot \text{LAB} - \text{vEastroof} \cdot \text{LAB}$$

$$\text{Normal} = 533.333 \, \text{lbf} \qquad \text{TENSION}$$

Figure 7.27 shows the answer to Question 1, as well as the axial force diagram of the strut being studied.

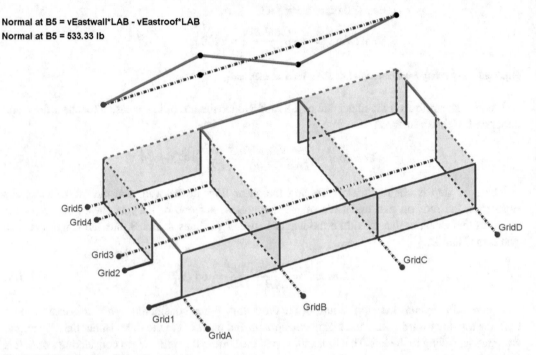

Fig. 7.27 Solution to question Question 1

Question 2 For eastern wind load only, what is the magnitude of the axial force in the strut along Grid1?

Question 3 For eastern wind load only, what is the magnitude of the axial force in the strut along Grid5 at the C5 intersection?

Question 4 For northern wind load only, what is the shear on the left side and the right side of the diaphragm between Gridline A and Gridline B?

Question 5 For northern wind load only, what is v_{wall} and what is v_{roof} along Gridline A?

Question 6 For northern wind load only, what is v_{wall} and what is v_{roof} along Gridline D?

Question 7 For northern wind load only, what is axial load in the strut along Gridline A at the intersection A2?

Question 8 For northern wind load only, what is axial load in the strut along Gridline D at the intersection D3?

Question 9 For northern wind load only, what is axial load in the strut along Gridline D at the intersection D4?

Answers to these questions are found in the following figures, but of course, it is best to attempt to answer all of the questions prior to looking at the solutions. All calculations could be done in a digital environment. The 3D images that follow were created as output from a parametric program written in the children's free software called GeoGebra.

The answer to Question 2 is zero. The strut between the diaphragm and the wall has no role to play, as there is no gap in the wall along gridline 1.

The answer to Question 3 is found in Fig. 7.28.

Fig. 7.28 Solution to question Question 3

Question 3

$$\text{Normal} + \text{vEastroof} \cdot (\text{LAB} + \text{LBC}) - \text{vEastwall} \cdot (\text{LAB}) = 0$$

$$\text{Normal} := \text{vEastwall} \cdot \text{LAB} - \text{vEastroof} \cdot (\text{LAB} + \text{LBC})$$

$$\text{Normal} = -533.333 \cdot \text{lbf} \qquad \text{COMPRESSION}$$

The answer to Question 4 is found in Fig. 7.29.

Question 4

the diaphragm flexes as a simply supported beam between A and B

$$\text{NorthABForce} := \text{wnorth} \cdot (\text{LAB})$$

$$\text{NorthABForce} = 10000 \cdot \text{lbf}$$

$$\text{VA} := \frac{\text{NorthABForce}}{2} = 5000 \cdot \text{lbf}$$

Fig. 7.29 Solution to question Question 4

The answer to Question 5 is found in Fig. 7.30.

Fig. 7.30 Solution to question Question 5

Question 5

$$\text{vAroof} := \frac{\text{VA}}{(\text{L}12 + \text{L}23 + \text{L}34 + \text{L}45)} = 125 \cdot \frac{\text{lbf}}{\text{ft}}$$

$$\text{vAwall} := \frac{\text{VA}}{(\text{L}12 + \text{L}23 + \text{L}34 + \text{L}45) - \text{L}12} = 200 \cdot \frac{\text{lbf}}{\text{ft}}$$

The answer to Question 6 is found in Fig. 7.31.

Fig. 7.31 Solution to
question Question 6

Question
6

$$\text{NorthCDForce} := \text{wnorth} \cdot (\text{LCD})$$

$$\text{NorthCDForce} = 10000 \cdot \text{lbf}$$

$$\text{VD} := \frac{\text{NorthCDForce}}{2} = 5000 \cdot \text{lbf}$$

$$\text{vDroof} := \frac{\text{VD}}{(\text{L12} + \text{L23} + \text{L34} + \text{L45})} = 125 \cdot \frac{\text{lbf}}{\text{ft}}$$

$$\text{vDwall} := \frac{\text{VD}}{(\text{L12} + \text{L23} + \text{L34} + \text{L45}) - \text{L34}} = 200 \cdot \frac{\text{lbf}}{\text{ft}}$$

The answer to Question 7 is found in Fig. 7.32.

Fig. 7.32 Solution to
question Question 7

Question
7
Since there is no wall under the strut, the easiest FBD is to go
from south to north

$$\text{Normal} := -1 \cdot \text{vAroof} \cdot \text{L12} = -1875 \cdot \text{lbf} \qquad \textbf{COMPRESSION}$$

The answer to Question 8 is found in Fig. 7.33.

Fig. 7.33 Solution to
question Question 8

Question 8

$$\text{Normal} + \text{vDroof} \cdot (\text{L12} + \text{L23}) - \text{vDwall} \cdot (\text{L12} + \text{L23}) = 0$$

$$\text{Normal} := \text{vDwall} \cdot (\text{L12} + \text{L23}) - \text{vDroof} \cdot (\text{L12} + \text{L23})$$

$$\text{Normal} = 1500 \cdot \text{lbf} \qquad \textbf{TENSION}$$

The answer to Question 9 is found in Fig. 7.34.

Question
9

Normal + vDroof·(L12 + L23 + L34) − vDwall·(L12 + L23) = 0

Normal := vDwall·(L12 + L23) − vDroof·(L12 + L23 + L34)

Normal = −375·lbf   COMPRESSION

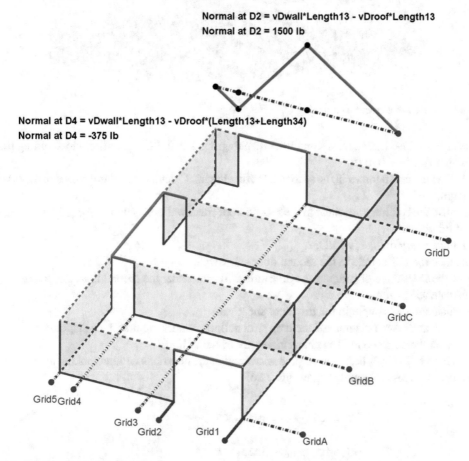

Normal at D2 = vDwall*Length13 - vDroof*Length13
Normal at D2 = 1500 lb

Normal at D4 = vDwall*Length13 - vDroof*(Length13+Length34)
Normal at D4 = -375 lb

GridD

GridC

GridB

Grid5 Grid4

Grid3

Grid2   Grid1

GridA

Fig. 7.34 Solution to question Question 9

Project 7–2 A two story building is subjected to known wind loads and is shown in Fig. 7.35. The width of the windward face (perpendicular to the wind) is 60 ft. between Gridlines 1 and 2. The length of the building in the direction of the wind is 40 ft. along Gridlines A, B, C. There are symmetric staggered shear walls on either end of the building. The width of the upper shear wall is LAB = 8 ft., the width of the lower shear wall is LBC = 10 ft. A continuous column (chord) runs over two stories on the windward side of the LFRS. A drag strut collector runs under the diaphragm, but above the shear walls the entire 40 ft. length. Continuous chords run along the entire 60 ft. length. The height of each floor is 12 ft.

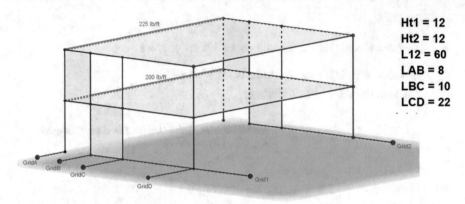

Ht1 = 12
Ht2 = 12
L12 = 60
LAB = 8
LBC = 10
LCD = 22

Fig. 7.35 Project 7–2 geometry and spacing

The uniform load hitting the second floor diaphragm is 225 lb./ft. The uniform load hitting the first floor diaphragm is 200 lb./ft.

Calculate the chord forces in the second floor diaphragm. Calculate the chord forces in the first floor diaphragm.

Calculate the equilibrating shear in the second floor shear walls. Calculate the equilibrating shear in the first floor shear walls.

Calculate *vdiaphragm2, vwall2*.

Calculate the *C/T* couple in the second floor shear wall.

Draw the axial force diagram of the drag strut collector under the second floor diaphragm.

Calculate *vdiaphragm1, vwall1*.

Calculate the *C/T* couple in the first floor shear wall.

Draw the axial force diagram of the drag strut collector under the first floor diaphragm.

Start with the easiest part. The chord forces are trivial to find.

As in Project 7–1, it is best to attempt to answer all of the questions without looking at the solutions. The chord forces are established in Fig. 7.36.

$$plf2 := 225 \; \frac{lbf}{ft} \qquad plf1 := 200 \; \frac{lbf}{ft}$$

$$Ht1 := 12ft \qquad\qquad Ht2 := 12ft$$

$$L12 := 60ft$$

$$LAB := 8ft \qquad\qquad LBC := 10ft \qquad\qquad LCD := 22ft$$

$$Mpeak2 := \frac{plf2 \cdot L12^2}{8} = 101250 \; lbf \cdot ft$$

$$Chord2 := \frac{Mpeak2}{LAB + LBC + LCD} = 2531.25 \; lbf \cdot ft$$

$$Mpeak1 := \frac{plf1 \cdot L12^2}{8} = 90000 \; lbf \cdot ft$$

$$Chord1 := \frac{Mpeak1}{LAB + LBC + LCD} = 2250 \; lbf$$

Fig. 7.36 Chord forces in Project 7–2

Now look at the second floor diaphragm. Engineers start with load flow from the top and move towards the foundation. Note that drag struts or collectors always run under the full length of diaphragms. That is why they are there, they connect to shear walls eventually. They also become chords when the loads are switched to a different direction. Find the *vdiaphragm* of the second floor and of the first floor. Find *vwall* of the second floor. Find *vwall* of the first floor but be careful! What force does the first floor really feel?

These unit shears, also known as shear flows, are shown in Fig. 7.37.

$$\text{Reaction2} := \frac{\text{plf2} \cdot \text{L12}}{2} = 6750 \text{ lbf}$$

$$\text{Reaction1} := \frac{\text{plf1} \cdot \text{L12}}{2} = 6000 \text{ lbf}$$

$$\text{vdiaph2} := \frac{\text{Reaction2}}{\text{LAB} + \text{LBC} + \text{LCD}} = 168.75 \frac{\text{lbf}}{\text{ft}}$$

$$\text{vdiaph1} := \frac{\text{Reaction1}}{\text{LAB} + \text{LBC} + \text{LCD}} = 150 \frac{\text{lbf}}{\text{ft}}$$

$$\text{vwall2} := \frac{\text{Reaction2}}{\text{LAB}} = 843.75 \frac{\text{lbf}}{\text{ft}}$$

$$\text{vwall1} := \frac{\text{Reaction1} + \text{Reaction2}}{\text{LBC}} = 1275 \frac{\text{lbf}}{\text{ft}}$$

Fig. 7.37 Unit shear flows in Project 7–2

Find the worst axial force in the drag strut (collector) under the second floor diaphragm. This will occur right where the wall ends. Figure 7.38 shows a sketch explaining the statics at that critical point.

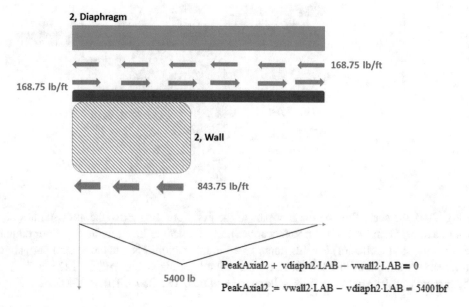

Fig. 7.38 Axial force in drag strut/collector

Now, find the *C T* couple at the bottom of the second floor wall. This is shown in Fig. 7.39

$$Reaction2 \cdot ht2 - CT2 \cdot LAB = 0$$

$$CT2 := \frac{Reaction2 \cdot Ht2}{LAB} = 10125 \text{ lbf}$$

Fig. 7.39 C T Couple calculation

Next, look at the statics of the drag strut (collector) under the first floor diaphragm. Always take a free body diagram of the drag strut to find its axial force. Figure 7.40 is a sketch explaining the statics.

Fig. 7.40 Shear flow on first floor diaphragm

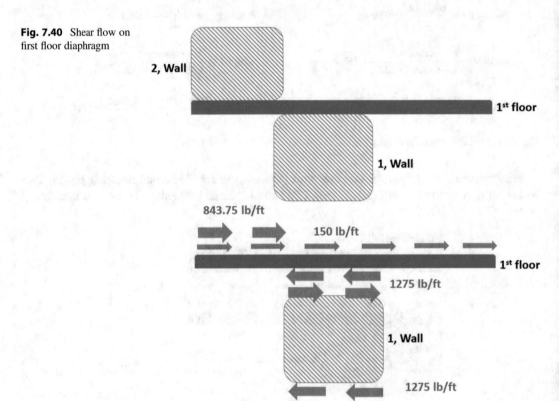

In Fig. 7.40, the shear flow on the first part of the first floor drag strut (the left part) has both the 843 lb./ft. coming from the second floor shear wall, and the 150 lb./ft. from the first floor diaphragm. Notice the drag strut (collector) is being compressed in this region. Then in the second part, above the first floor shear wall, it feels the 150 lb./ft. from the first floor diaphragm and the 1275 lb./ft. from the first floor shear wall. The net shear flow is the superposition of two flows. This is described in Fig. 7.41.

$$OnDragPart1 := vwall2 + vdiaph1 = 993.75 \frac{lbf}{ft}$$

$$OnDragPart2 := vdiaph1 - vwall1 = -1125 \frac{lbf}{ft}$$

Fig. 7.41 Net shear flow calculations

The tensile peak force is found at the end of the second floor shear wall.

$$Tensile\ Peak = OnDragPart1 \cdot LAB = 7950\ lb \qquad (7.6)$$

The compressive peak is next found. Start at positive 7950 lb and move to this new peak point.

$$Compressive\ Peak = Tensile\ Peak + OnDragPart2 \cdot LBC = -3300\ lb \qquad (7.7)$$

The axial force diagram is shown in Fig. 7.42.

Fig. 7.42 Axial force
diagram Project 7–2

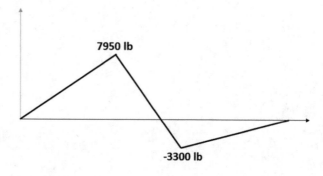

Rigid Diaphragms

8

Walls in buildings are constructed of various types and materials. They may be organized as exterior vs interior, or as load bearing vs non-load bearing, or as prefabricated vs built-in-place. Any walls which are classified as load bearing means that they carry vertical (downward) floor loads, as well as lateral loads. Walls are typically planar, but they may also be designed with flanges, or as a closed cell assembly such as the elevator core of a tall building. Exterior walls may be load bearing, or they may only be required to act as a protective environmental shield. If acting as shields, such walls are known as "envelopes" or non load bearing "curtain walls". Even if a wall is a curtain wall, it must carry its own vertical self weight safely and it must be able to transfer lateral wind loads to the building support structure. These cladding structures have tributary areas, part of which transfers loads to other building elements and eventually these loads move down to the ground. A common and economical way of transferring lateral load down to the ground is via shear walls.

Shear Walls

In a building, the vertical structural planes, called walls, are tied together by the horizontal planes which are called diaphragms. Lateral forces from seismic events are applied directly to these horizontal diaphragm elements. It is important to fully understand the behavior of vertical walls, which are rigidly attached to the ground, but cantilever up from the ground and connect to the horizontal diaphragms. These walls, known as shear walls, carry the lateral loads down to the ground. They can be arranged many different ways in a building, yet the arrangement of the walls must address programmatic issues, which means the aesthetics and the functionality or practicality of the placement. These walls can be reinforced masonry walls made of brick or concrete block, or they can be cast in place concrete walls, or they can be precast concrete walls which are erected on site.

The wall proportion height/width, also known as the "aspect ratio", will have a profound effect on the behavior of the wall. Introductory architectural engineering classes have always dealt with slender members, so imagine the following wall which is still relatively slender in its own plane. This wall is fixed at its base, free at its top and it is subjected to a lateral load at the top only. This walls greatly exaggerated deformed shape is shown in Fig. 8.1.

© Springer Nature Switzerland AG 2020
E. Saliklis, *Structures: A Studio Approach*, https://doi.org/10.1007/978-3-030-33153-5_8

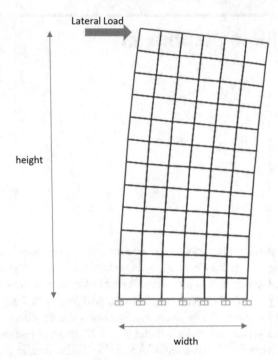

Fig. 8.1 Cantilever shear wall deformation greatly exaggerated

Note how the lateral load is applied directly to the top of the shear wall, this is where the horizontal floor or diaphragm would be. Also note how its deformed shape is what one would expect from a cantilever, albeit a vertical cantilever.

In this type of shear wall, traditional undergraduate formulas for stress apply because the deformation is what is typically expected. A subtle but very important point is that axial stress induced by this kind of bending is vertical, not horizontal! In the cantilever wall of Fig. 8.1, the axial stress would be tensile on the left side and compressive on the right side for the load shown. The aspect ratio, (height/width) ratio is 2. For slender shear walls with an aspect ratio greater than 3 all of traditional undergraduate beam formulas apply.

Since this shear wall is a traditional cantilever, there would be no tensile stress at the vertical centerline of the wall because that would align with the neutral axis.

Next consider a wall that is "long" i.e. the height/width ratio is smaller than 1. If the height/width ratio is approximately is 1 or lower, then the wall behaves very differently than our slender cantilever. The primary deformation is a shear distortion, also known as "racking". The classic shear racking deformed shape is shown in Fig. 8.2. It is not the traditional cantilever type of bending.

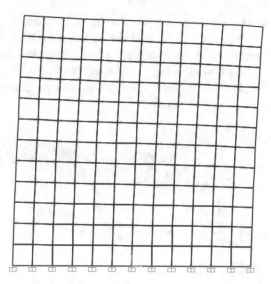

Fig. 8.2 Shear racking deformed shape of shear wall

Compressive stresses and shear stresses in shear walls are not going to be a cause for concern, they will be nowhere near the allowable stress. A dangerous stress for concrete, i.e. the stress that might cause concrete to fail in compression could be on the order of 3000 lb/in^2 or 20 MPa. Compressive stresses in shear walls will not approach these values. But note that while concrete is an excellent material to take compressive stress (either induced by axial compression or by flexure as in the side of a vertical cantilever beam), it is extremely weak in tension. Reinforcement must be placed in all concrete structures as the direction of the loads constantly changes, one side or the other of a cantilever could experience tension. Building codes use the symbol fc' to denote this allowable compressive stress. Recall that Codes used the letter f_b denote bending stress, whereas mechanics textbooks used σ. Yet axial stress is always force /length^2, regardless of whether it arises from axial load or from bending. But the allowable stress for concrete always uses the letter fc'. High strength concrete can have fc' values as high as 10,000 psi (69 MPa) today.

What about the modulus of elasticity E? That does not capture the strength of a material, as does fc', rather it describes stiffness.

Rigid Diaphragm Supported by Symmetrically Placed Shear Walls

A horizontal element in a structure which experiences in-plane, not out-of-plane loads is called a diaphragm. This can be a floor, or a roof. There are two types of assumptions used when analyzing diaphragms, they can be consider as either rigid or flexible. Note: if the diaphragm is considered rigid, the elements supporting the diaphragm, such as shear walls, are considered to be flexible. If the diaphragm is considered to be flexible, then the supporting elements are assumed to be rigid. Such binary classifications are perfectly acceptable and they greatly simplify the analysis. For example, if a roof diaphragm is considered to be rigid, then the entire roof plane translates and rotates as a single (rigid) entity. The shear wall previously shown in Fig. 8.1 was assumed flexible, it bent as a cantilever, thus the load at the top of the wall would have arisen from a rigid roof diaphragm. Regardless of whether such a laterally loaded shear wall deforms either primarily in flexure as a cantilever (narrow wall), or primarily in shear as a deep beam, where the "usual" flexural rules no longer apply (long wall), the binary classification still calls for the rigid roof and flexible wall assumption. Next, consider the stiffness of a structural member which is its resistance to deformation. Clearly some shear walls are stiffer than others. In Fig. 8.3, a horizontal rigid diaphragm (roof plate) is supported by shear walls which must be flexible. The planes of the shear walls line up with the global y axis. The z axis is up.

Fig. 8.3 Rigid diaphragm supported by four flexible shear walls

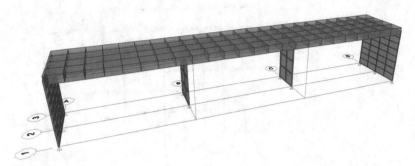

Now suppose that this rigid roof diaphragm was taking loads in the y direction (note there are no lateral load carrying members in the x direction!). All four walls will act to resist this lateral load, but will the shorter interior walls take the same amount of force as the full exterior walls, or will they take more, or will they take less? To answer the question, imagine that the rigid diaphragm roof is supported by four lateral springs. Two springs have $k_{outside}$ and two springs have k_{inside}. The assumption that the roof diaphragm is rigid requires that all four springs deform the same lateral amount. But stiffness is defined as:

$$stiffness = k = \frac{force}{length} \tag{8.1}$$

And

$$force = k \cdot \Delta = \frac{force}{length} \cdot length \tag{8.2}$$

If the displacement in each of the four resisting springs is the same, it makes perfect sense that the force in each wall (spring) is proportional to the stiffness of that wall (spring). Thus, the stiffer wall (spring) will pick up more load than the less stiff wall.

Figure 8.4 shows the previous building with a greatly exaggerated lateral deformation.

Fig. 8.4 Rigid diaphragm
greatly exaggerated lateral
deformation

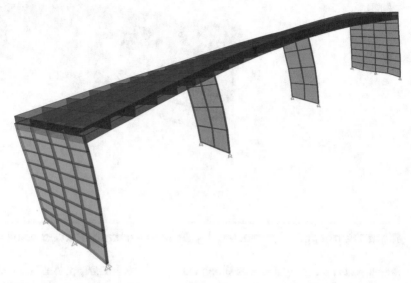

It follows that if the stiffness of the outside walls is twice that of the interior walls, then the outside walls would take *2 force* and the inside walls would take *1 force* which means the total original lateral load must be divided in to *6 force*.

Thus for some $P_{lateral}$ applied to the rigid roof, for equilibrium along the y axis we have

$$P_{lateral} = 2F + F + F + 2F \tag{8.3}$$

Note that although there are four equilibrating forces here, the problem still is statically determinate. This is so, because this technique only applies to a rigid roof, wherein each wall deforms the same amount, as was imagined in the spring analogy. This allowed for the force in each wall to be found as a function of its stiffness. But how is stiffness calculated? Of course there are structural engineering formulas to calculate actual stiffness, but it is best to consider the relative stiffness of each wall, i.e. how stiff is each wall compared to its neighbors.

The following formula does just that, it shows that for walls that bend more than they rack, made of identical material, and with equal heights (z dimension) the distribution of forces is:

$$P_{lateral} = P_{lateral}\left(\frac{I_1}{\sum I_i}\right) + P_{lateral}\left(\frac{I_2}{\sum I_i}\right) + P_{lateral}\left(\frac{I_3}{\sum I_i}\right) + P_{lateral}\left(\frac{I_4}{\sum I_i}\right) \tag{8.4}$$

or

$$P_{lateral} = P_{wall1} + P_{wall2} + P_{wall3} + P_{wall4} \tag{8.5}$$

where I_i is the individual moment of inertia of wall$_i$ and $\sum I_i$ is the summation of all the moment of inertias. This equation clearly shows how the total lateral load is divided up into percentages, which must total 100% of load, in proportion to the stiffness of each wall.

For example, as shown in Fig. 8.5, if all four walls were the same size, then all the moment of inertia I values would be identical, thus each wall would take ¼ of the total load. This is a testament to the linkage between structural form and structural forces. Decisions about form should not be made arbitrarily because they have profound structural implications!

Fig. 8.5 Force in each wall
is proportional to its
stiffness

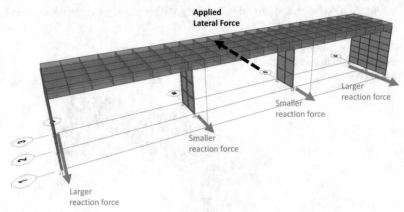

Rigid Diaphragm Supported by Asymmetrically Placed Shear Walls

Suppose next that a rigid roof is supported by an unsymmetric placement of shear walls. Such a roof
would not experience pure translation, there would have to be rotation plus translation. Why? For a
symmetric arrangement of shear walls supporting a rigid roof, with the roof translating the same
amount throughout, the stiffness of each laterally supporting element (spring) determines how much
load flows to it. Stiffer elements pick up more load. Yet in Fig. 8.6, a horizontal rigid diaphragm (roof
plate) is supported by asymmetrically placed flexible shear walls.

Fig. 8.6 Rigid diaphragm
supported laterally by
asymmetrically placed
flexible shear walls

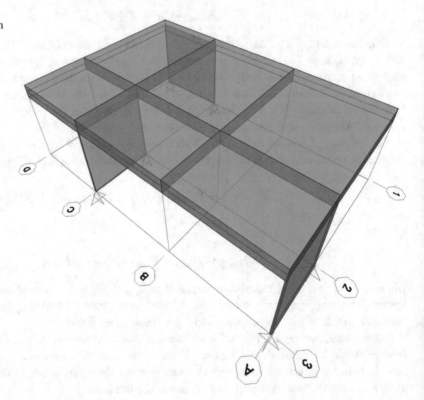

In Fig. 8.7, the shear walls are on grid lines A and C, between 3 2 and 3, and on grid line 1 between C and D. For such a configuration of LFRSs, the seismic load either in the x or in the y direction, will induce torsion of the roof as well as translation. This is because the seismic roof load is passed through the Center of Mass (CM) of the roof, as Force = mass·acceleration.

When performing the statics by hand, one must consider the intersecting shear walls as separate, distinct units. There is no shear "flow" around the corner of intersecting walls. Also, assume the shear force in each wall to be constant, simply label it as F1, F2, F3. And if there are any vertical columns holding up the roof (there are none shown in Fig. 8.7), neglect them in these calculations. Vertical columns, even those with fixed connections, cannot provide lateral stability in buildings in California. There are three unknown LFRS forces, thus the problem is statically determinate.

Fig. 8.7 Three shear walls means statically determinate configuration

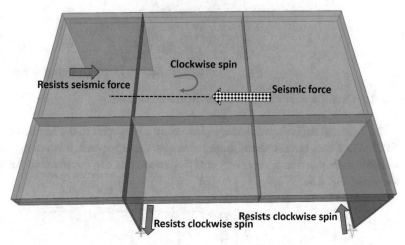

It has already been noted that cantilever action of a shear wall means that pure bending exists, and that lateral movement of the wall is dictated by the moment of inertia. The stiffness of a cantilever with moment of inertia I, and height h is that is completely free at the top is

$$k_{cantilever} = \frac{3 \cdot E \cdot I}{height^3} \tag{8.6}$$

Such a cantilever is shown once again in Fig. 8.8. But this time, the flow nets of force are shown. Blue contours depict compressive flow of force and green contours represent tension. The density of the contours reflect the magnitude of the forces.

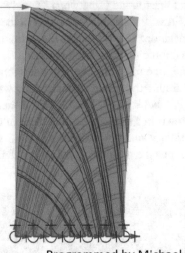

Programmed by Michael Goldenberg

Fig. 8.8 Cantilever shear wall flow nets shown as contour lines

In structural analysis problems, an important counterpart to the cantilever is a member that is fully fixed at one end, but restrained from rotation at the other end. It is not restrained against translation, imagine the top carries a very, very stiff diaphragm that can translate but not rotate. Such a member is shown in Fig. 8.9.

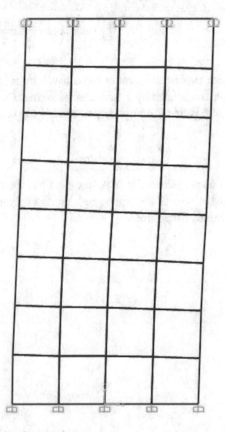

Fig. 8.9 Shear with top restrained against rotation

The stiffness of such a structure that is restrained from rotation on top is:

$$k_{fixed\ fixed} = \frac{12EI}{height^3} \tag{8.7}$$

The addition of rotational restraint at the top of the structure of Fig. 8.9 makes it four times stiffer than the wall of Fig. 8.8. Another way of thinking about it is that more energy is required to move such a wall compared to one that is free on top. If there are multiple shear walls in one direction, they act as a group of springs supporting a rigid diaphragm, and they all contribute to resisting the movement of a rigid, (inflexible) diaphragm.

Now consider a different type of distortion, one that is not governed by cantilever bending, but is governed by shear. The dashed line in Fig. 8.10 shows where the wall was originally, and the solid blue lines show straight (or nearly straight) shear deformation, i.e. the deformation here arises primarily from shear, not from bending. The material property G, known as the shear modulus, will influence the deformation, and the moment of inertia I will not play a role at all.

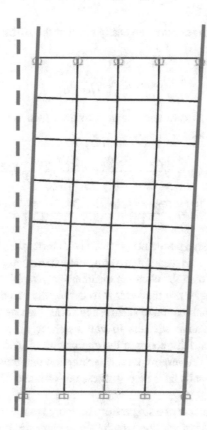

Fig. 8.10 Shear deformation does not depend on moment of inertia

The formula for shear deformation of such a structure is:

$$\Delta_{shear} = \frac{constant \cdot Force \cdot height}{A \cdot G} \tag{8.8}$$

For a rectangular cross section, the constant, known as a "shape factor" is 1.2. Although it is not strictly correct to do so, we make the assumption that reinforced concrete is an isotropic materials thus:

$$G = \frac{E}{2 \cdot (1 + \nu)} \tag{8.9}$$

and the Poisson's Ratio ν for concrete and masonry makes the relationship between G and E roughly

$$G \approx 0.4E \tag{8.10}$$

Thus the previous force displacement relationship becomes:

$$\Delta_{shear} = \frac{3 \cdot Force \cdot height}{A \cdot E} \tag{8.11}$$

Using the previous idea of stiffness, we can easily see that the stiffness of a shear wall exhibiting only shear deformation or racking is:

$$k_{shear\ only} = \frac{A\,E}{3\,height} \tag{8.12}$$

If a shear wall exhibits both flexure and shear distortion, then it is necessary to superpose the previous stiffness equations to find:

$$\Delta_{free\ on\ top} = \frac{Force \cdot height^3}{3EI} + \frac{3 \cdot Force \cdot height}{A \cdot E} \tag{8.13}$$

$$\Delta_{fixed\ fixed} = \frac{Force \cdot height^3}{12EI} + \frac{3 \cdot Force \cdot height}{A \cdot E} \tag{8.14}$$

Here is a subtlety that might have gone undetected. The deformation actually increases because of the additional shear deformation. This means that a structure experiencing shear deformation as well as bending deformation is less stiff than the classical, bending-only cantilever.

Structural engineers often simply use the letter R to denote the value of the relative stiffness of a wall. The stiffness arises from combination of flexure and shear, but one or the other could be ignored based on office practice. The relative stiffness of various shear walls becomes important because torsion will exist if there is eccentricity between the application of load and the center of resistance, i.e. the "centroids of the springs". A very important assumption that must be used when studying the interaction of walls is that shear walls have zero stiffness out-of-plane. Only in-plane stiffness is used to resist lateral loads.

In Fig. 8.11, a qualitative plan view of a single story building is shown. Four shear walls of uniform thickness, material and height, but of varying lengths are to carry the lateral load. In this picture, the center of mass (CM) of the roof is at the centroid of the rectangular roof because of symmetry and the weight of the walls is ignored, again this is common practice. But the LFRS is clearly asymmetric because of the various lengths of the walls.

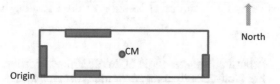

Fig. 8.11 Qualitative plan view of four shear walls

Earthquake lateral loads are always, always passed through the center of mass (CM). This is true because these inertial forces arise from mass multiplied by acceleration (F = m·a), and the mass is idealized as being condensed down to a single point. Suppose the previous building was subjected to some lateral E/W earthquake load passing through the CM. This is shown in Fig. 8.12. It is visually obvious that the center of rigidity (CR), is somewhat north of the CM because the wall on the north side is longer than the wall on the south side. Since the north wall is stiffer, the entire diaphragm would rotate counter-clockwise about the CR.

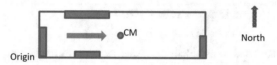

Fig. 8.12 Load passes through CM but roof spins about CR

While the approximate location of the CR was simply intuited in Fig. 8.12, the equations to calculate the stiffness of each wall can be used for quantitative evaluation of where the CR actually lies. The quantitative technique finds each of the four different stiffnesses, and then they are compared to, and scaled by the smallest (or the largest) value. Say the south wall has the lowest stiffness, it can be assigned a value of R = 1. The east wall may be R = 1.5 or whatever the k_{east}/k_{south} value actually is. The west wall may be R = 2 and the north wall may be R = 2.5 for example, depending on the actual wall stiffness values.

To illustrate the technique, those four R values will be used for a building that is subjected to two earthquake loads, one going East and one going North. Seismic loads always pass through the CM. Actual lengths will allow for the quantitative location of the CR as shown in Fig. 8.13.

Fig. 8.13 Quantitative plan view of four shear walls

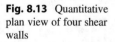
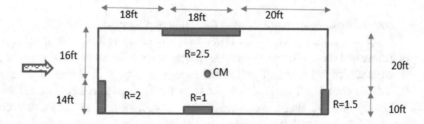

The calculations used to establish the exact location of the CR along the x axis are shown in Fig. 8.14

Fig. 8.14 Calculating eccentricity in x

Arbitrarily set the origin at the southwest corner.
For North/South loads, only the west wall and the east wall resist loads.

$$\text{Rsouth} := 1 \qquad \text{Rwest} := 2 \qquad \text{Rnorth} := 2.5 \qquad \text{Reast} := 1.5$$

$$\text{xwest} := 0\text{ft} \qquad \text{xeast} := 18\text{ft} + 18\text{ft} + 20\text{ft}$$

$$\text{xbar} := \frac{\text{Rwest} \cdot \text{xwest} + \text{Reast} \cdot \text{xeast}}{\text{Rwest} + \text{Reast}} = 24\,\text{ft}$$

$$\text{xCM} := \frac{18\text{ft} + 18\text{ft} + 20\text{ft}}{2} = 28\,\text{ft}$$

$$\text{eccentricityx} := \text{xCM} - \text{xbar} = 4\,\text{ft}$$

Similar calculations are used to establish the exact location of the CR along the y axis. These are shown in Fig. 8.15

Fig. 8.15 Calculating eccentricity in y

Only the north wall and the south wall resist east/west loads

$$\text{ysouth} := 0\text{ft} \qquad \text{ynorth} := 10\text{ft} + 20\text{ft}$$

$$\text{ybar} := \frac{\text{Rsouth} \cdot \text{ysouth} + \text{Rnorth} \cdot \text{ynorth}}{\text{Rsouth} + \text{Rnorth}} = 21.429\text{ft}$$

$$\text{yCM} := \frac{10\text{ft} + 20\text{ft}}{2} = 15\,\text{ft}$$

$$\text{eccentricityy} := \text{yCM} - \text{ybar} = -6.429\text{ft}$$

Both of these eccentricities match the intuitive placement of the CR.

In summary, the relative stiffness of a wall is denoted by R, and the R value could be a combination of flexure and shear stiffness. Assuming that the least stiff wall has an R value of 1.0, a wall that is three times as stiff as this least stiff wall would have an R value of 3. The calculation of the x and y values of the CR are completely analogous to the steps for finding the centroid of a shape.

Finally, the Building Code requires the incorporation of an accidental eccentricity, even if the structural layout is such that zero eccentricity was planned. The purpose of this provision is to account for errors in construction and to allow for movement of significant dead load items. ASCE 7 Section 12.8.4.2 requires an eccentricity of 5% of the dimension perpendicular to the applied force. So, for an East/West earthquake, eccentricity would be, at a minimum, 5% of the North/South dimension. Note: the eccentricity is used to make the wall effect worse than it really is. The Code requires the use of this extra 5% length to either subtract less in the direction of spinning, or to add more in the direction of spinning.

The superposition of translation of a rigid roof and rotation about its CR is aided by the very important concept of General Plane Motion (GPM). This concept allows for the analysis of one complicated but true scenario, by studying two separate but false scenarios! Step 1 is to assume (falsely) that pure translation exists. This analysis assumes that each of the resisting shear walls (springs) experience the same amount of movement. This Step 1 analysis always neglects the out-of-plane walls. Step 2 is to assume (falsely) that pure rotation exists about the CR. All shear walls will resist this motion. Finally, Step 1 is superposed to Step 2. The superposition may be additive for some walls, and subtractive for other walls, i.e. the resisting reaction in a particular wall may be in one direction for the Step 1 translation analysis, but it may be in the opposite direction for the Step 2 rotation analysis.

Figure 8.16 examines the pure translation scenario a bit closer. There are four shear walls in total, one each edge of the building which is shown in plan view. Yet for a northward seismic load, only the West and East walls resist shear, the North and South walls do nothing since the load is perpendicular to their planes. Note this is Step 1 of a two-step process and as such, it temporarily ignores torsion, i.e. it is pure translation. Thus, it really does not matter where the load is applied, this will be a pure translation assumption.

Fig. 8.16 Step 1, assume pure translation

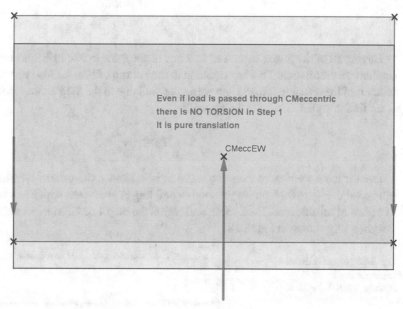

Even if load is passed through CMeccentric
there is NO TORSION in Step 1
It is pure translation

CMeccEW

If two walls resist shear, the shear must be divided in proportion to the two resisting elements' relative stiffness. Then, for the walls in Fig. 8.16 the first step of GPM is shown in Fig. 8.17

Fig. 8.17 Shear is divided
in proportion to relative
stiffness, shown
algebraically

$$\Delta = \Delta West = \Delta East$$

all the deformations are identical in pure translation

$$\Delta = \frac{FWest}{kWest} = \frac{FEast}{kEast}$$

arbitrarily choose to express all the forces in terms of FWest

$$V = BaseShear = FWest + FEast$$

$$V = FWest + \frac{FWest \cdot kEast}{kWest}$$

simplifying a bit gives $V = \frac{FWest}{kWest} \cdot (kWest + kEast)$

$$V = \frac{FWest}{kWest} \cdot \Sigma k$$

Having all of the forces expressed in terms of the force in one LFRS is convenient. It allows for an equilibrium calculation. This last equation at the bottom of Fig. 8.17 is very versatile. It allows for the solution of the resisting force in any particular wall due to the Step 1 assumption of pure translation. It is not tied to wall 1. Thus for the ith wall:

$$F_i = \frac{V \cdot k_i}{\sum k_i} \tag{8.15}$$

Even if there were more than two walls resisting shear, the pattern established by Eq. 8.15 would still be valid. Figure 8.18 shows yet another building in plan view, which has three shear walls resisting a northward seismic load. Each shear wall resists the Step 1 of GPM in accordance to its stiffness. This is stylistically shown in Fig. 8.18.

Fig. 8.18 Shear is divided
in proportion to relative
stiffness, shown graphically

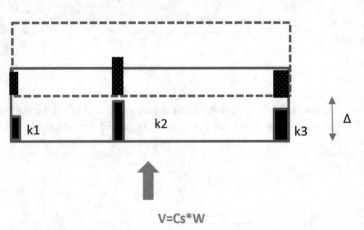

Returning to the building in Fig. 8.16 with four shear walls, Step 2 can now be studied since the first step of the General Plane Motion analysis has been completed. Step 2 is the assumed case of pure torsion about the CR, no translation at all.

In Fig. 8.19, there are four walls that act as the LFRS. The southern wall has a $R = 1$ value (relative stiffness). The West, North and East walls all have an $R = 4$ relative stiffness. Assume that the seismic force is going north through an eccentric CM which is placed to the right of the CR. Thus, counter-clockwise spinning will ensue. The premise of Step 2 of General Plane Motion is that pure torsion will exist and spinning occurs about the CR. Notice in Fig. 8.19, the distances from each wall to the CR. Also notice in Fig. 8.19 that for a counter-clockwise seismic moment, all four walls kick in to resist the motion by inducing a net clockwise spin. But how to find these forces?

Fig. 8.19 Step 2, assume pure rotation

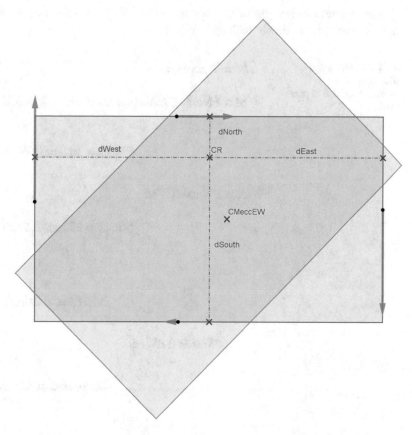

Recall that arc length is:

$$s = r \cdot \theta \tag{8.16}$$

The most challenging part of the analysis is the next step. The North Wall moves west through its arc length but it does not move north or south. The East Wall moves north through its arc length but it does not move east or west. The South Wall moves east through its arc length but it does not move north or south. And the West Wall moves south through its arc length, but it does not move east or west. There is a small paradox here, one of the few paradoxes you will encounter in the undergraduate curriculum! The paradox is that we assume that when a point (wall) swings through some angle, it translates "sideways" but it does not get "pulled in closer". There is only one rotation θ but there are four different arc lengths. Thus, the movement of the North Wall 1 is dNorth·θ and movement of East Wall 2 is dEast·θ, the movement of the South Wall is dSouth·θ and the movement of the West Wall is dWest·θ There is another surprise here that may seem paradoxical, the forces are placed in their original spots on the roof! These force were shown in Fig. 8.19 and they act in the opposite direction of the movement of the wall, as they are induced by this movement. The force in each wall is a product of its stiffness multiplied by its arc length movement.

Moment equilibrium can now be established. The applied seismic moment is M, this is the Base Shear V applied at some eccentricity to the CR.

$$M = V \cdot eccentricity_x \tag{8.17}$$

A counter-clockwise spin of the roof induces a clockwise resistance from all four walls. These induced moments are slowly built up in Fig. 8.20.

Fig. 8.20 First steps in calculating wall displacements due to pure rotation

$$M = V \cdot eccentricityx$$

$$M = FNorth \cdot dNorth + FEast \cdot dEast + FSouth \cdot dSouth + FWest \cdot dWest$$

$$FNorth = kNorth \cdot \Delta North$$

$$FWest = kWest \cdot \Delta West \qquad FEast = kEast \cdot \Delta East$$

$$FSouth = kSouth \cdot \Delta South$$

$$\Delta North = \theta \cdot dNorth$$

$$\Delta West = \theta \cdot dWest \qquad \Delta East = \theta \cdot dEast$$

$$\Delta South = \theta \cdot dSouth$$

As was done in the Step 1 pure translation analysis, all forces will be expressed in terms of some arbitrarily chosen one, and as before, all forces in Step 2 will be expressed in terms of FWest. The common feature between all terms will be the rotation θ (Fig. 8.21).

Fig. 8.21 Calculating wall forces due to pure rotation

$$\theta = \frac{\Delta \text{North}}{d\text{North}} = \frac{\Delta \text{East}}{d\text{East}} = \frac{\Delta \text{South}}{d\text{South}} = \frac{\Delta \text{West}}{d\text{West}}$$

$$\theta = \frac{\dfrac{F\text{North}}{k\text{North}}}{d\text{North}} = \frac{\dfrac{F\text{East}}{k\text{East}}}{d\text{East}} = \frac{\dfrac{F\text{South}}{k\text{South}}}{d\text{South}} = \frac{\dfrac{F\text{West}}{k\text{West}}}{d\text{West}}$$

a small simplification leads to:

$$\frac{F\text{North}}{k\text{North} \cdot d\text{North}} = \frac{F\text{East}}{k\text{East} \cdot d\text{East}} = \frac{F\text{South}}{k\text{South} \cdot d\text{South}} = \frac{F\text{West}}{k\text{West} \cdot d\text{West}}$$

thus

$$F\text{North} = \frac{F\text{West} \cdot k\text{North} \cdot d\text{North}}{k\text{West} \cdot d\text{West}}$$

$$F\text{West}$$

$$F\text{East} = \frac{F\text{West} \cdot k\text{East} \cdot d\text{East}}{k\text{West} \cdot d\text{West}}$$

$$F\text{South} = \frac{F\text{West} \cdot k\text{South} \cdot d\text{South}}{k\text{West} \cdot d\text{West}}$$

Plugging these expressions for all the forces back into the moment equilibrium equation gives the following steps shown in Fig. 8.22.

Fig. 8.22 Moment equilibrium with four shear walls

$$M = V \cdot \text{eccentricityx}$$

$$M = F\text{North} \cdot d\text{North} + F\text{East} \cdot d\text{East} + F\text{South} \cdot d\text{South} + F\text{West} \cdot d\text{West}$$

$$M = \frac{F\text{West} \cdot k\text{North} \cdot d\text{North}}{k\text{West} \cdot d\text{West}} \cdot d\text{North} + \frac{F\text{West} \cdot k\text{East} \cdot d\text{East}}{k\text{West} \cdot d\text{West}} \cdot d\text{East} \ldots$$
$$+ \frac{F\text{West} \cdot k\text{South} \cdot d\text{South}}{k\text{West} \cdot d\text{West}} \cdot d\text{South} + F\text{West} \cdot d\text{West}$$

To find the unknown resisting force FWest, first factor it out. Then, factor out the kWest·dWest term from the denominators. These two steps are shown in Fig. 8.23.

Fig. 8.23 Extracting a
single wall force from the
moment equation

$$V \cdot eccentricityx = FWest \cdot \left(\begin{array}{l} \dfrac{kNorth \cdot dNorth}{kWest \cdot dWest} \cdot dNorth + \dfrac{kEast \cdot dEast}{kWest \cdot dWest} \cdot dEast \ldots \\[2ex] + \dfrac{kSouth \cdot dSouth}{kWest \cdot dWest} \cdot dSouth + dWest \end{array} \right)$$

$$V \cdot eccentricityx = \dfrac{FWest}{kWest \cdot dWest} \cdot \left(\begin{array}{l} kNorth \cdot dNorth \cdot dNorth + kEast \cdot dEast \cdot dEast \ldots \\ + kSouth \cdot dSouth \cdot dSouth + dWest \cdot kWest \cdot dWest \end{array} \right)$$

There was nothing special about the West Wall. This technique could have been used for any wall i. Then, the conclusion is to use Eq. 8.18 for any wall in the Step 2 portion of GPM.

$$F_i = \frac{External\ Moment \cdot (k_i\ d_i)}{\sum (k_i \cdot d_i^2)} \tag{8.18}$$

Having concluded Step 2, Pure Rotation, of the General Plane Motion Problem, allows for the superposition of each wall's resisting force, that from Step 1 and that from Step 2.

The Direct Stiffness Method

The previous problem can be solved using the direct stiffness method. This method is used heavily in Matrix Methods structural analysis courses. It is also used in Structural Dynamics when calculating the stiffness of a rigid roof supported by lateral force resisting elements.

An example begins with Fig. 8.24. This particular example was designed by Professor Graham Archer. The goal of this exercise is to calculate the load flowing into the *south lateral force resisting system* due to an Earthquake Load in the *Eastward Direction*. The plan of the building is 105 × 60 units of length. Assume that the LFRS are as shown, two act in the East/West direction (stiffness is 4000 force/length and 1000 force/length for these) and two act in the North/South direction (stiffness is 4000 force/length for each of these). The Center of Mass of the uniform deck slab is found by inspection to be at the center of the plan. Assume the seismic weight W is 666666.6 units of force and assume that Cs = 0.15. Again, the seismic event throws the roof to the right. Figure 8.24 shows the original CM and it also shows the moved CM which adds the Code mandated eccentricity. The new eccentricity between CMeccentric and CR is 21 ft. The reason the shift of the CM was to the south was that if the seismic load spins the roof, the seismic moment will be increased by this shift, since moments are always summed about the CR and in this example, the CR is certainly north of the center of the roof. If we were seeking the force in the north wall for a seismic load to the right, the shift of the CM would be to the north because this would lessen the reduction of force induced by the moment as the force in the north wall goes to the left due to pure translation and to the right to resist seismic moment. The Code asks the designer to study each LFRS separately, and to always inflate its reaction, or to lessen its reduction, by means of this additional eccentricity.

Fig. 8.24 Geometry in plan view of direct stiffness method example

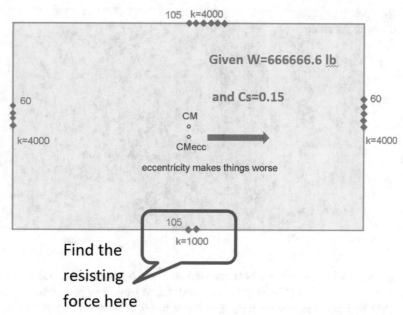

The base shear V is quickly found.

$$V = C_s \cdot W = 0.15 \cdot 666666.6lb = 100000lb \tag{8.19}$$

The Code calls for moving the Center of Mass based on 5% of the axis length perpendicular to the force. Since the force is in the east direction, the eccentricity can shift the CM either north or south, whatever makes the problem worse. Recall that this problem is exclusively calculating the force in the southern LFRS wall. It is clear that the CM can be moved 5% of 60 units of length.

$$Code\ eccentricity\ y = 0.05 \cdot 60ft = 3ft \tag{8.20}$$

Note that in Step 2 of GPM, the distance between the CMeccentric and the CR will be needed as the "arm" for the Seismic Moment. This arm will be 21 ft. and is shown in Fig. 8.25.

In Step 1 of GMP, since the force in the south wall is sought, and since the pure translation induces a south wall force to the left, we seek an increase of moment which for the shift to be southward.

The x value of center of rigidity is found by inspection along the East/West axis because of the symmetry of the East and West walls. The y value of the center of rigidity must be calculated. This can be done algebraically or graphically. A graphical solution is shown in Fig. 8.25, using the Inverse Axis Method.

Fig. 8.25 Y location of
CR found graphically

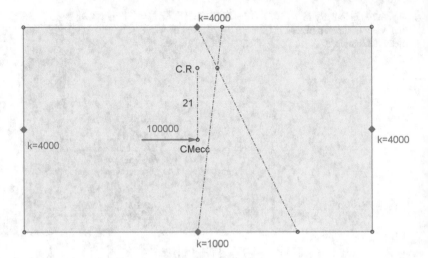

Step 1 will be to assume pure translation. There is no spinning, and the force in the North Wall and the force in the South Wall is immediately found from the proportions of their stiffnesses, 4/5 to the north and 1/5 to the south. Here, the diaphragm is SOLELY PUSHED in the East Direction, through the CR. If there are more than two walls resisting shear, the shear is divided in proportion to each of the resisting elements' relative stiffness. Figure 8.26 shows the proportioning due to pure translation.

Fig. 8.26 Forces due to
Step 1 pure translation
found graphically

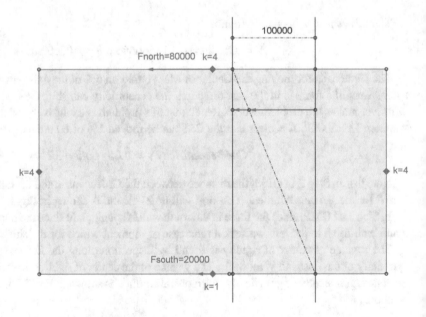

To ensure that the original seismic load, which passed through the eccentric CM is statically equivalent to the situation where the load is passed through the CR, a moment must be added. The magnitude of this moment is

$$Meccentric = 100000 \cdot 21 = 2100000 force * length \qquad (8.21)$$

Where the 21 ft. arm was previously established, this is the distance between the CMeccentric and the CR. Fig. 8.27 shows the Seismic Moment.

Fig. 8.27 Equivalent force and moment about CR

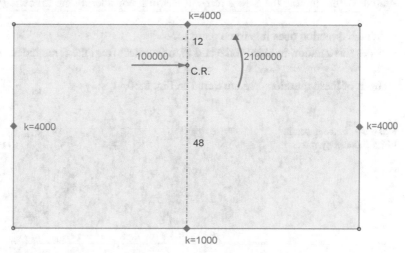

The forces of the LRFSs resist the moment induced by a unit rotation of the slab. This moment is readily found as:

$$M_{unit\ \theta} = \sum k_i \cdot d_i^2 \qquad (8.22)$$

Figure 8.28 shows how the moment induced by a unit θ is obtained.

$$\text{Munit } \theta := 2 \cdot \left[4000 \frac{lbf}{ft} \cdot (52.5ft)^2 \right] + 4000 \frac{lbf}{ft} \cdot (12ft)^2 + 1000 \frac{lbf}{ft} \cdot (48ft)^2$$

$$\text{Munit } \theta = 24930000 lbf \cdot ft$$

Fig. 8.28 Moment induced by a unit θ

Clearly, a unit rotation is much greater than the spin induced by the Seismic Moment. Calculate the angle that matches the Seismic Moment. Figure 8.29 shows how the the actual θ which is induced by the Seismic Moment is obtained.

$$\text{Mseismic} := 21ft \cdot 100000 lbf$$

$$\theta := \frac{\text{Mseismic}}{\text{Munit } \theta} = 0.084$$

Fig. 8.29 Actual θ induced by the Seismic Moment

This nearly completes the static equilibrium approach to this problem. The force induced in the South LRFS is 20,000 lbf due to Step 1, the Pure Translation part of General Plane Motion. This was calculated from a proportioning of spring stiffness. Now, what is the force induced by Pure Rotation? Note that the magnitude of the rotation is now known to be 0.084 radians. The last step is surprisingly simple. Before finding the Step 2 force, it is worth considering the following qualitative issues:

- Which direction does this force point to?
- Does this rotation-induced force add to, or subtract from the force induced by translation?

  Both of these questions are answered in Fig. 8.30.

Fig. 8.30 Force in south LFRS induced by rotation

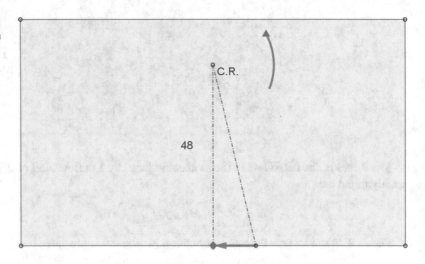

Fsouth from rotation= θ * radius * ksouth
Fsouth rot = 0.084 *48 * 1000
Fsouth rot = 4043.52

Having the θ immediately allows for the calculation of the arc length. Here the radius is calculated from the CR to the south wall, i.e. 48 ft. Having the arc length, the force in the south wall is found since force is equal to stiffness multiplied by the displacement, which is the arc length.

$$F_{south\ total} = 20000lb + 4043lb = 24043lb \tag{8.23}$$

This calculation will be repeated, but it will be found in an entirely different manner. This time, the same exercise will be performed using the Direct Stiffness Method. This method is very powerful and useful, but at first glance it may seem complicated. It isn't. To use this method:

- Assign two translational and one rotational degree of freedom to the roof. The choice of where to place the rotational degree of freedom is arbitrary. If the rotational degree of freedom is placed at the CR (Center of Rigidity), then the stiffness matrix will be diagonal. This is helpful for inversion of the stiffness matrix.
- One unit degree of freedom will be imposed at a time, with the other two degrees of freedom forced to be zero. This will create three distinct scenarios. For each scenario, quickly establish the forces

induced in each LRFS due the imposition of the one single unit DOF imposition. Consider these as "internal forces" which arise in response to the imposition of movement or rotation.
- Calculate what is the magnitude of each external equilibrating force in each DOF. There will be three such responses, two translational and one rotational, for each imposition. These form the three columns individually of the stiffness matrix, with a total of nine entries.

Begin by re-drawing the roof plan of Fig. 8.24, but this time add in three degrees of freedom. In this example they are chosen to be placed at the CM. This is shown in Fig. 8.31.

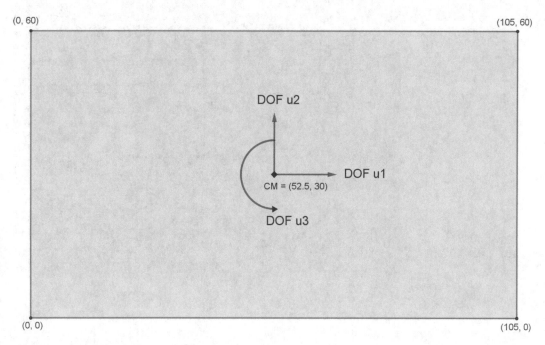

Fig. 8.31 Three degrees of freedom are placed at the CM

But the load is passed through the CMeccentric which is 3 ft. South. This is shown in Fig. 8.32.

Fig. 8.32 Load is passed
through the CM eccentric

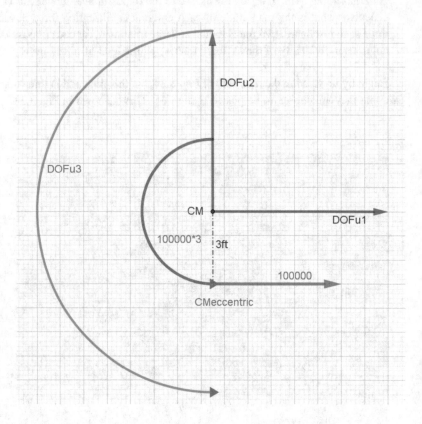

Thus the force vector is formed, with the first item describing the externally applied force in the u1 direction, the second item describing the external force in the u2 direction and the third item describing the applied seismic moment which is in the u3 direction.

$$ForceVector = \left\{ \begin{array}{c} 10000 \\ 0 \\ 10000 \cdot 3 \end{array} \right\} = \left\{ \begin{array}{c} 10000 \\ 0 \\ 30000 \end{array} \right\} \tag{8.24}$$

Next, impose a unit displacement along the first DOF, ensuring that the other two DOF remain zero. This means the entire roof shifts 1 unit of length to the right. Note the DOF remain at the CM even though the loads are applied to the CMeccentric. This is shown in Fig. 8.33.

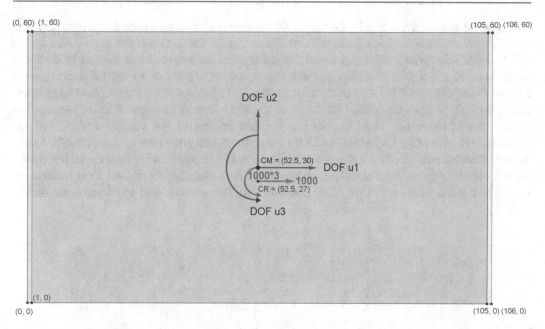

Fig. 8.33 DOF do not change position, though loads are applied to CMeccentric

Establish the forces induced in the LFRS due to this unit movement. Think of them as "internal loads" that develop due to the straining of the restraints. Since there is no north/south movement, the LFRSs aligned along the north/south experience no strain and thus develop no force. The LFRSs aligned along the east/west respond to the unit movement in proportion to their stiffness. This is shown in Fig. 8.34.

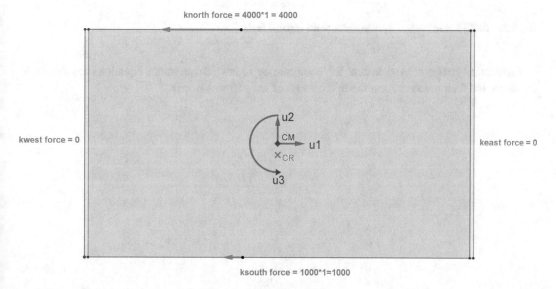

Fig. 8.34 Forces induced only in LFRSs aligned along the east/west

If students are confused at all by the Direct Stiffness method, it often occurs at this next step. The next step will establish the first column of the Stiffness Matrix. The entries of this first column are the reactions that equilibrate the internal forces just found from the imposition of the first DOF. These reactions occur at the point of the degrees of freedom, not at the point of the applied external loads. And the "internal forces" that need equilibrating are at the LFRS, not at the point of load application. Recall that this problem established the DOFs at the CM, whereas the external loads were applied through the CMeccentric. Thus, to establish the first column of the stiffness matrix, find the equilibrating forces at the CM where the DOFs were placed, and place each of those equilibrators in its spot in the first column of the stiffness matrix. These unknown equilibrating forces, which will form the first column of the stiffness matrix, are shown in Fig. 8.35. Note that the original force vector is not needed here and is not shown in Fig. 8.35, only the LFRS forces which need equilibrating are shown.

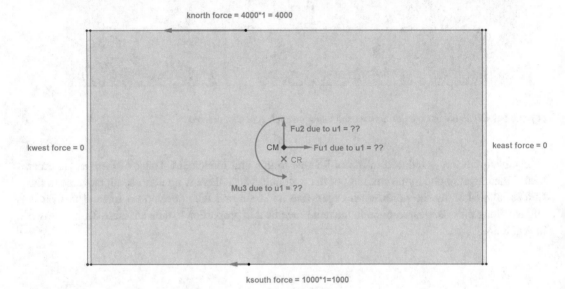

Fig. 8.35 LFRS forces must be equilibrated, not the external loads

Elementary statics is used to find the equilibrating forces. These statics equations are shown in Fig. 8.36 and their answers form the first column of the stiffness matrix.

$$\Sigma Fx = 0 \qquad\qquad\qquad \Sigma Fy = 0$$

$$Fu1 - 4000 - 1000 = 0 \qquad\qquad Fu2 := 0$$

$$Fu1 := 4000 + 1000 = 5000$$

$$\Sigma MatCM = 0$$

$$Mu3 + 4000 \cdot 30 - 1000 \cdot 30 = 0$$

$$Mu3 := -4000 \cdot 30 + 1000 \cdot 30 = -90000$$

$$K := \begin{pmatrix} 5000 & \blacksquare & \blacksquare \\ 0 & \blacksquare & \blacksquare \\ -90000 & 0 & \blacksquare \end{pmatrix}$$

Fig. 8.36 Elementary statics to find the equilibrating force

Next, impose a unit movement in the u2 DOF. Ensure DOF1 = DOF3 = 0. Quickly note what forces are induced in the LRFS by this move. Call these "internal forces" and these arise from strains in the LFRS. The u2 movement and the induced internal forces are shown in Fig. 8.37.

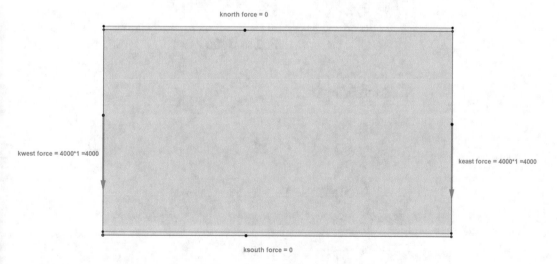

knorth force = 0

kwest force = 4000*1 =4000

keast force = 4000*1 =4000

ksouth force = 0

Fig. 8.37 Only u2 movement is non-zero and the corresponding internal forces

Now establish the second column of the Stiffness Matrix. The entries of this second column are the reactions that equilibrate the internal forces just found from the imposition of the second DOF. Again, these reactions occur at the chosen point of the degrees of freedom. These unknowns are shown in Fig. 8.38. Note that as always in the Direct Stiffness Method, the forces that need equilibrating arise at the LFRS, from the imposition of the constrained unit movement.

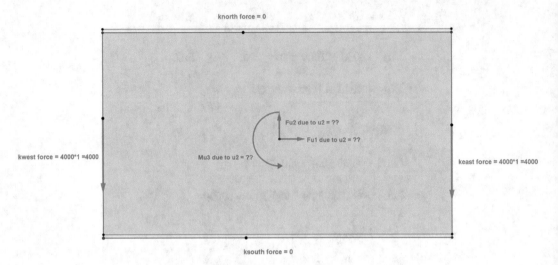

Fig. 8.38 Forces that need equilibrating arise at the LFRS due to u2

Elementary statics is used to find these equilibrating forces, which then are placed into the second column of the stiffness matrix, as shown in Fig. 8.39.

$$\Sigma Fx = 0 \qquad\qquad \Sigma Fy = 0$$

$$Fu1 := 0 \qquad\qquad Fu2 := 4000 + 4000$$

$$Fu2 = 8000$$

$$\Sigma MatCM = 0$$

$$Mu3 := 0$$

$$K := \begin{pmatrix} 5000 & 0 & \blacksquare \\ 0 & 8000 & \blacksquare \\ -90000 & 0 & \blacksquare \end{pmatrix}$$

Fig. 8.39 Form second column of stiffness matrix through elementary statics

Finally, impose a unit third DOF. This is u3 and it is a counterclockwise rotation since we use the right hand rule when defining rotations. This u3 rotation is shown in Fig. 8.40.

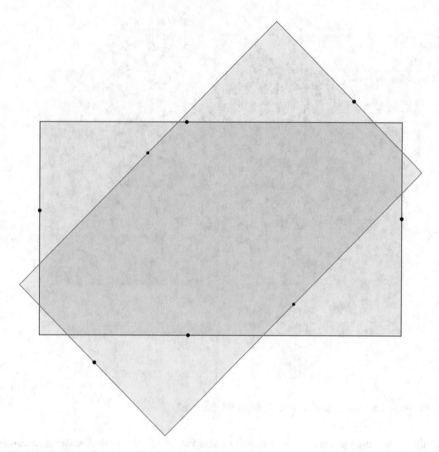

Fig. 8.40 Impose u3 = 1 while u1 = u2 = 0

Next establish the "internal" forces which develop because of this rotation. Notice that each force is $F = k \cdot d$ and d the distance is $d = r \cdot \theta$, thus $d = r$, because $\theta = 1$. Also notice that if the roof spins counterclockwise in line with the third DOF, then each LRFS resists it by spinning clockwise. Here is one more mini-paradox. The angle is assumed to be small but we show it as $45°$ to ensure that $s = r$ since $\theta = 1$. When swinging through the arc, the movement is perpendicular to the arc length, there is no movement at all along the radius of the arc, it does not draw nearer to the center of curvature of the arc, only perpendicular. Place the forces that develop in their original location in the plan view.

Due to their relatively larger size, the forces developed due to the unit u3 DOF rotation will be drawn to a different scale than the forces due to u1 and u2. These u3 DOF forces are shown in Fig. 8.41.

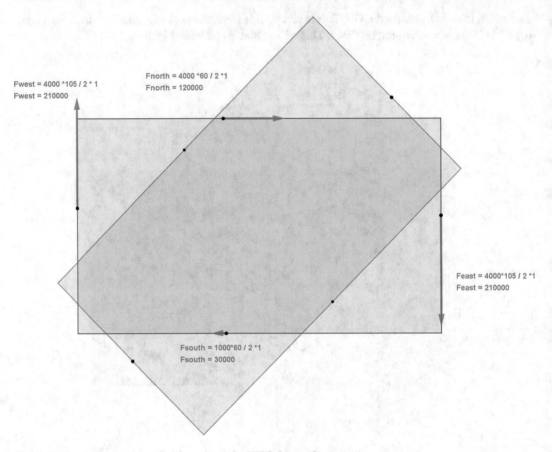

Fig. 8.41 Forces that need equilibrating arise at the LFRS due to u3

Notice the wide range of force values developed in Fig. 8.41, and note that these forces are much larger than the forces developed by the imposition of the translation DOFs.

The last column of the stiffness matrix is found by solving for the equilibrating two forces and one moment aligned with the original DOF, which were arbitrarily placed at the CM. These, as yet unknown equilibrators, are shown in Fig. 8.42.

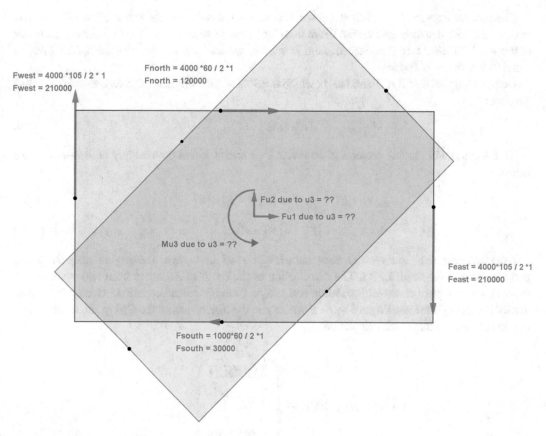

Fig. 8.42 Equilibrators due to u3 are at DOF

The equilibrators are found via statics yet again, forming the third column of the stiffness matrix as shown in Fig. 8.43.

$$\Sigma Fx = 0 \qquad\qquad \Sigma Fy = 0$$

$$120000 - 30000 + Fu1 = 0 \qquad\qquad Fu2 := 0$$

$$Fu1 := -90000$$

$$\Sigma MatCM = 0$$

$$Mu3 - 2 \cdot 210000 \cdot 52.5 - 120000 \cdot 30 - 30000 \cdot 30 = 0$$

$$Mu3 := 26550000$$

$$K := \begin{pmatrix} 5000 & 0 & -90000 \\ 0 & 8000 & 0 \\ -90000 & 0 & 26550000 \end{pmatrix}$$

Fig. 8.43 Elementary statics to find the equilibrating force due to u3

The entire stiffness matrix K is now known. It is a 3×3. From Physics we know $\{F\} = [K] \cdot \{d\}$ and we seek the 3×1 displacement vector. From that displacement vector we will extract the displacement of the south LFRS. Once that displacement is known, we can find the force in that LFRS since its individual stiffness is known.

Considering the overall stiffness matrix which is a 3×3 and the original force vector which is a 3×1 we have:

$$\{F\} = [K] \cdot \{d\} \tag{8.25}$$

This method will give us vector d if we invert 3×3 matrix K and pre-multiply both sides by 3×3 matrix K^{-1}

$$[K]^{-1} \cdot \{F\} = [K]^{-1} \cdot [K] \cdot \{d\} \tag{8.26}$$

$$[K]^{-1} \cdot \{F\} = 1 \cdot d \tag{8.27}$$

Note that obtaining the d vector means that $u1$, $u2$ and $u3$ are known. The original question was to find the force in the south LFRS. The deformation of the south LFRS arises from two sources, first from u1 which is pure east/west translation and second from u3 which is rotation, which will become translation through the arc length $s = r \cdot \theta$ where r is the distance from the CM to the south LFRS. Figure 8.44 shows this quick calculation.

$$\text{ForceVector} = \begin{pmatrix} 100000 \\ 0 \\ 300000 \end{pmatrix}$$

$$\text{displVector} := K^{-1} \cdot \text{ForceVector}$$

$$\text{displVector} = \begin{pmatrix} 21.516 \\ 0 \\ 0.084 \end{pmatrix}$$

Fig. 8.44 Obtain displacement vector

So $u1$ is 21.516 and $u3$ is 0.084. This is very good news, since in Fig. 8.29 the rotation was found to be 0.084 via the other method.

How to translate this back to a force in the South LFRS?

The force in the South LFRS is the superposition of the displacement due to u1 and the $s = r \cdot \theta$ displacement due to rotation u3, all multiplied by $[K]$ of the LFRS.

$$FsouthLFRS = 1000 \cdot (u1 + 30 \cdot u3) \tag{8.28}$$

Where 1000 is the lateral stiffness of the south LFRS (lb/ft) and 30 is the distance (ft) from the CM to the south LFRS.

$$FsouthLFRS = \frac{1000lb}{ft} \cdot (21.516ft + 30ft \cdot 0.0842rad) = 24042 \; lb \tag{8.29}$$

This exactly matches the force in the south LFRS that was found previously.

To really drive home the technique of the Direct Stiffness Method, the problem could be re-solved with a new choice for the placement of the Degrees of Freedom at the point of the south LFRS. This would be advantageous because rotation would not affect the problem the roof would assume to spin about the south LFRS. Thus, only u1 would be needed to find the force in the south LFRS. The stiffness matrix K would change as would the external force vector F. The new DOF are shown in Fig. 8.45, and the three required shifts along the newly placed DOF which are shown in Fig. 8.46.

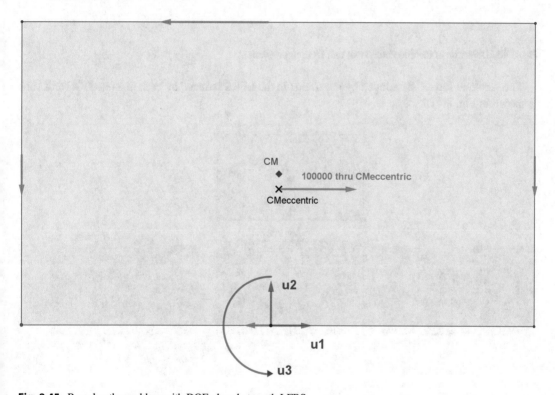

Fig. 8.45 Re-solve the problem with DOF placed at south LFRS

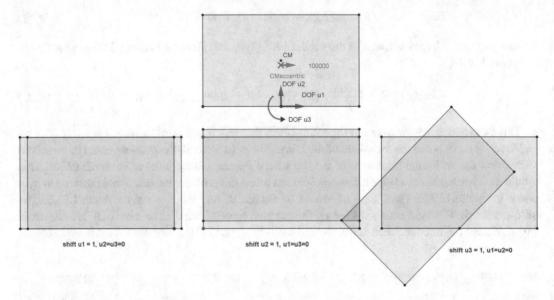

Fig. 8.46 Depiction of the three individual unit DOF impositions

The "internal forces" developed by the strains in the LFRS induced by each discrete DOF unit shift is shown in Fig. 8.47.

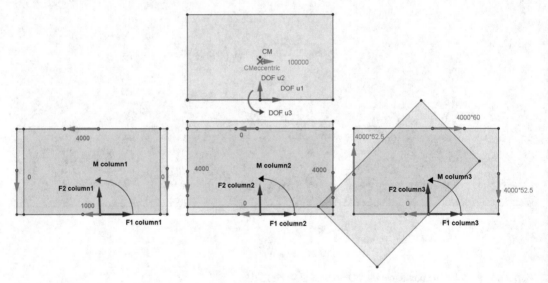

Fig. 8.47 Each unit imposition requires three equilibrators at DOF

Figure 8.48 shows the internal forces at the LFRS due to a unit u1 = 1 and u2 = u3 = 0. These forces must be equilibrated by forces at the chosen DOF, here at the south LFRS. These equilibrators then form the first column of the stiffness matrix.

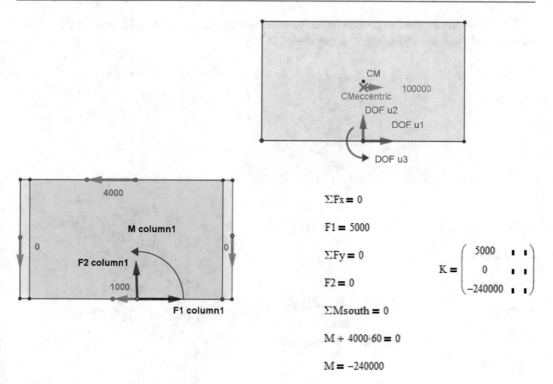

$$\Sigma Fx = 0$$

$$F1 = 5000$$

$$\Sigma Fy = 0$$

$$F2 = 0$$

$$K = \begin{pmatrix} 5000 & \blacksquare & \blacksquare \\ 0 & \blacksquare & \blacksquare \\ -240000 & \blacksquare & \blacksquare \end{pmatrix}$$

$$\Sigma Msouth = 0$$

$$M + 4000 \cdot 60 = 0$$

$$M = -240000$$

Fig. 8.48 Solution for first three equilibrators

Similarly, Fig. 8.49 shows the equilibrating forces for the second shift, i.e. $u2 = 1$ and $u1 = u3 = 0$.

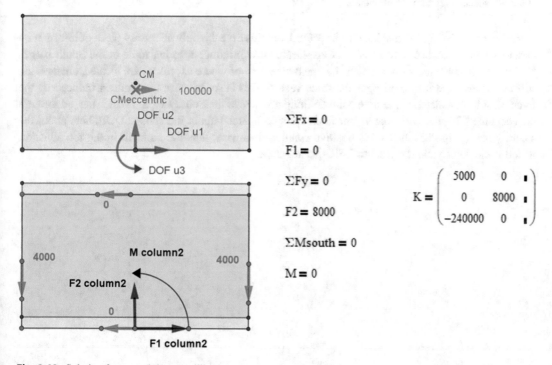

$$\Sigma Fx = 0$$

$$F1 = 0$$

$$\Sigma Fy = 0$$

$$F2 = 8000$$

$$K = \begin{pmatrix} 5000 & 0 & \blacksquare \\ 0 & 8000 & \blacksquare \\ -240000 & 0 & \blacksquare \end{pmatrix}$$

$$\Sigma Msouth = 0$$

$$M = 0$$

Fig. 8.49 Solution for second three equilibrators

And finally the third column of the stiffness matrix arises from the third shift, namely u3 = 1 and u1 = u2 = 0, with the equilibrating forces shown in Fig. 8.50.

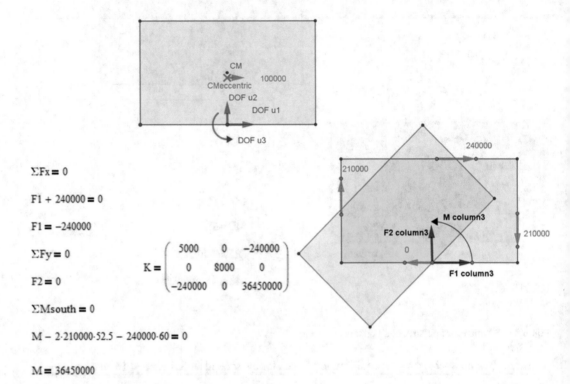

ΣFx = 0

F1 + 240000 = 0

F1 = −240000

ΣFy = 0

F2 = 0

$$K = \begin{pmatrix} 5000 & 0 & -240000 \\ 0 & 8000 & 0 \\ -240000 & 0 & 36450000 \end{pmatrix}$$

ΣMsouth = 0

M − 2·210000·52.5 − 240000·60 = 0

M = 36450000

Fig. 8.50 Solution for third three equilibrators

As before, solve for the displacement vector d, but this time the only necessary piece of information is the u1 value, because only East/West displacement will induce a spring force in the South LRFS. Notice that this new location of DOFs, i.e. the assignment of where u1, u2, u3 are located, changes the stiffness matrix and it also changes the force vector! This is so, because the moment induced by the force must be calculated at the new point defining u3, here at the South LRFS. Load is passed through the eccentric CM, but the force vector finds the spin induced about the chosen DOF, here about the south. Note in Fig. 8.51 that u3 is positive counter-clockwise, thus the moment in the force vector, which is clockwise about the south LFRS, is negative.

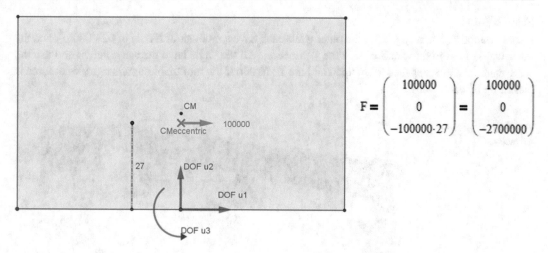

Fig. 8.51 Force vector is based on DOF

Having this new, alternate force vector, and the new, alternate stiffness matrix, both of which relate to the DOF being placed at the south LFRS, the new displacement vector can be found, identifying these alternate values of u1, u2 and u3. The steps needed to do this are shown in Fig. 8.52.

$$Falt := \begin{pmatrix} 100000 \\ 0 \\ -2700000 \end{pmatrix}$$

$$Kalt := \begin{pmatrix} 5000 & 0 & -240000 \\ 0 & 8000 & 0 \\ -240000 & 0 & 36450000 \end{pmatrix}$$

$$dalt := Kalt^{-1} \cdot Falt = \begin{pmatrix} 24.043 \\ 0 \\ 0.084 \end{pmatrix}$$

Fig. 8.52 Obtain displacement vector at south LFRS

And the force in the South LRFS is exactly as it was before:

$$F_{South} = k_{South} \cdot u_1 = 1000 \, ^{lb}\!/_{ft} \cdot 24.043 ft = 24043 \, lb$$

Notice that the rotation of the South wall (0.084) about the DOF at the South Wall, does not induce any force in the south wall. Only in-plane translation induces force.

Project 8–1

Part 1 Suppose that in Fig. 8.53, the outer walls each have a moment of inertia of 622,080in⁴, and the
inner walls are 10in thick and 6 ft. in the y direction. All the walls have the same height and use the
same material and are spaced 20 ft. apart and are 12 ft. tall. If the roof diaphragm is rigid, what force is
picked up by each wall?

Fig. 8.53 Project 8–1
geometry

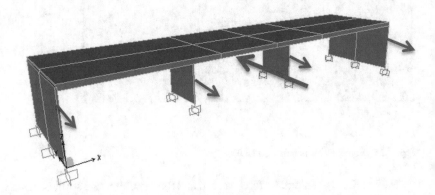

Part 2 Repeat part 1 but use SAP. Make the roof use a material that has a very large E, 10000 times
the E of concrete. Compare SAP results to hand calcs, either by adding up all the y forces of the pin
reactions at the base of each wall. That is the shear in each wall. Comment on the agreement.

Notice the title of this section used the word "symmetric". Suppose next that the rigid roof is
supported by an UNSYMMETRIC placement of shear walls. What do you expect to happen other than
pure translation of the rigid roof?

Part 3 Use SAP to create a new model, or simply modify your existing model of Problem 2, but this
time create an unsymmetric arrangement of 3 (only 3) shear walls supporting a rigid roof. Do not
corroborate your answers by hand, (unless you want to do so for fun). This exercise is meant to let you
explore and literally "play" a bit with the program. No need to print any pictures of the deformed shape
but do tell me what you see happening that is different than the symmetric case.

The solution to part1 is shown in Figs. 8.54 and 8.55

Problema 1

InertiaOuter := 622080

width := 6·12

Given

$$\text{InertiaOuter} = \frac{\text{thickouter} \cdot \text{width}^3}{12}$$

$+$

Find(thickouter) → 20

width := width·in = 6 ft

thickouter := 20in

$$\text{InertiaOuterCheck} := \frac{\text{thickouter} \cdot \text{width}^3}{12} = 622080 \, \text{in}^4$$

thickinner := 10in

$$\text{InertiaInner} := \frac{\text{thickinner} \cdot \text{width}^3}{12} = 311040 \, \text{in}^4$$

Fig. 8.54 Initial solution to part 1 of Project 8–1

Fig. 8.55 Completion of solution to part 1 of Project 8–1

$$\text{SumofInertias} := 2 \cdot \text{InertiaOuterCheck} + 2 \cdot \text{InertiaInner} = 1866240 \, \text{in}^4$$

V := 10000lbf

$$V = V \cdot \frac{\text{InertiaOuterCheck}}{\text{SumofInertias}} + V \cdot \frac{\text{InertiaInner}}{\text{SumofInertias}} + V \cdot \frac{\text{InertiaInner}}{\text{SumofInertias}} + V \cdot \frac{\text{InertiaOuterCheck}}{\text{SumofInertias}}$$

$$\frac{\text{InertiaOuterCheck}}{\text{SumofInertias}} = 0.333 \qquad \frac{\text{InertiaInner}}{\text{SumofInertias}} = 0.167$$

Outer walls will each take 1/3 of V, inner walls will each take 1/6 of V

Project 8–2 All shear walls are to be assumed as being pinned at their extreme lower corners. In Fig. 8.56, the shear walls are reinforced concrete, 1 ft. thick but ignore their weight and in SAP, release their out-of-plane shear carrying capabilities. The roof is considered as rigid. Make the roof 1 ft. thick steel solely to add mass and rigidity, we don't design 1 ft. thick steel plate floors in real life. There are two shear walls in the y direction and one shear wall in the x direction. Make all the columns pinned-pinned, i.e. release strong and weak moment at each end. Make the columns weightless steel.

Fig. 8.56 Project 8–2
geometry

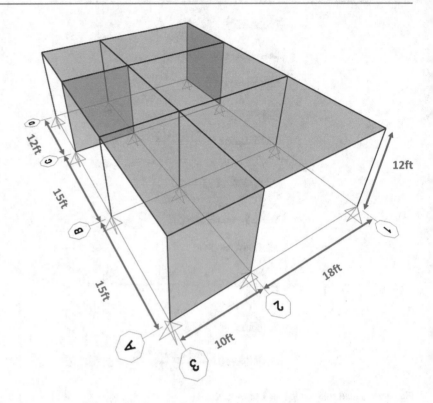

Seismic Weight is known. Cs = 0.3 in both directions. Find Shear wall forces. The details are shown in Fig. 8.56. Analyze the building in Fig. 8.56 by hand and by SAP. Snip output from SAP and take pictures of your hand calcs. Combine into a document and compare the results. Nothing fancy but keep it neat and tidy. Define two new load patterns, call one EQX for earthquake in the X direction and one EQY for earthquake in the Y direction. Touch all the roof elements and ASSIGN >> AREA LOADS >> GRAVITY then put in your GRAVITY MULTIPLIER which represents Cs. Here, use 0.3 g in the X direction for Load Pattern EQX and 0.3 g in the Y direction for Load Pattern EQY. This is shown in Fig. 8.57.

Fig. 8.57 Using SAP2000
to create static seismic load
in X

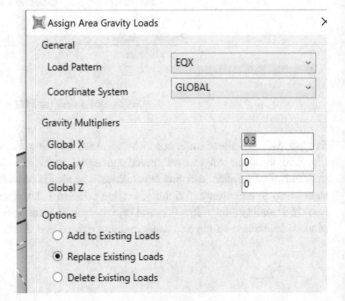

Be sure to throw the building in the X direction for EQX and in the Y direction for EQY. Fig. 8.58 shows the Object Load Assign for EQY

Fig. 8.58 Using SAP2000 to create static seismic load in Y

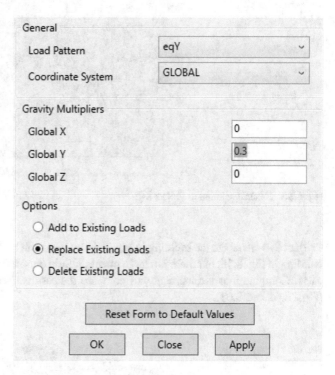

The solution to project 8–2 is shown in Figs. 8.59 and 8.60

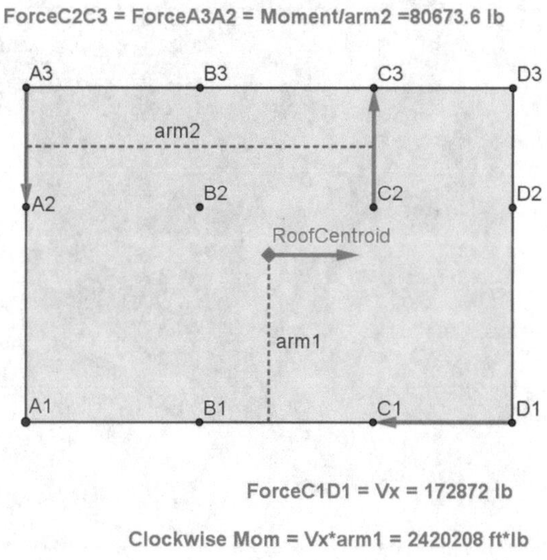

ForceC2C3 = ForceA3A2 = Moment/arm2 =80673.6 lb

ForceC1D1 = Vx = 172872 lb

Clockwise Mom = Vx*arm1 = 2420208 ft*lb

Fig. 8.59 Initial solution to Project 8–2

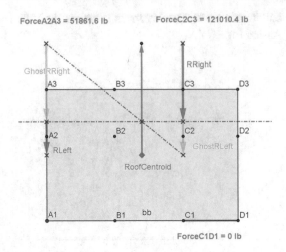

Fig. 8.60 Finished solution to Project 8–2

Project 8–3 Analyze the building in Fig. 8.61 by hand and by SAP. It has the same dimensions as the building in Fig. 8.56, but there is a large opening. Snip output from SAP and take pictures of your hand calcs. Combine into a document and compare the results. Nothing fancy but keep it neat and tidy (Figs. 8.62 and 8.63).

Fig. 8.61 Project 8–3 geometry

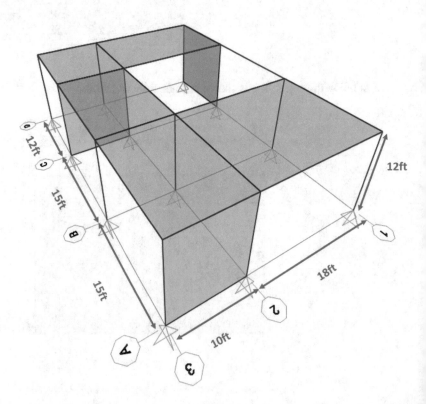

The solution to project 8–3 is shown in Figs. 8.60 and 8.61

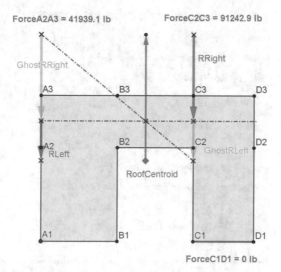

Fig. 8.62 Initial solution to Project 8–3

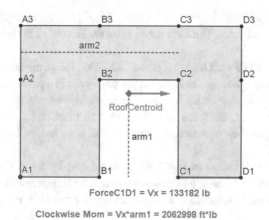

Fig. 8.63 Finished solution to Project 8–3

When solving this in SAP2000, don't forget to release the capabilities of the shear walls to resist out-of-plane forces. Remember, people cannot push over a shear wall out of plane, but an earthquake can. Figure 8.64 shows the details of how to do this, click on the shear wall, ASSIGN> > AREA> > EDGE RELEASES

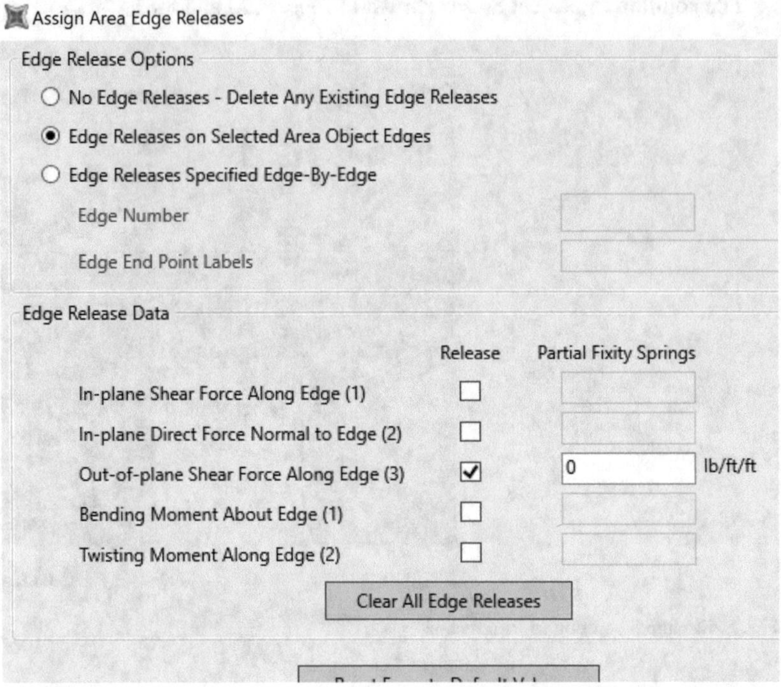

Fig. 8.64 Assigning edge
releases to walls in
SAP2000

Use DISPLAY> > SHOW TABLES > > STRUCTURE OUTPUT to get the BASE REACTIONS for your three different load cases. But then zoom in on the pins of each shear wall to verify your hand calculations.

The W (seismic weight) of your building is the GLOBAL FZ of the DEAD Case. Notice the out of plane forces in the torsional case sum to zero, but the pin forces in a particular wall are not zero! This is the huge, major insight of this entire lesson. There are two shear walls in Y direction, and for earthquake in the X direction, the force in each shear wall must matchy match to sum to zero since there is no external load in they Y direction, yet each wall does indeed have base shear!

Project 8–4 Analyze the building in Fig. 8.61 by hand and by SAP. But this time the dead load between lines C and D is twice the load elsewhere, as it is a corridor. Snip output from SAP and take pictures of your hand calcs. Combine into a document and compare the results. Nothing fancy but keep it neat and tidy.

The solution to project 8–4 is shown in Figs. 8.65 and 8.66

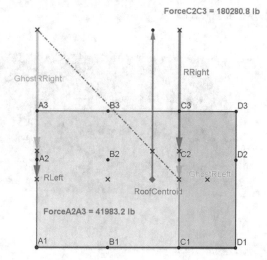

ForceC2C3 = 180280.8 lb

GhostRRight

RRight

ForceA2A3 = 41983.2 lb

ForceC1D1 = 0 lb

Fig. 8.65 Initial solution to Project 8–4

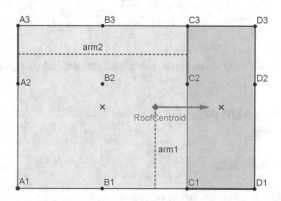

ForceC2C3 = ForceA3A2 = Moment/arm2 =103723.2 lb

ForceC1D1 = Vx = 222264 lb

Clockwise Mom = Vx*arm1 = 3111696 ft*lb

Fig. 8.66 Finished solution to Project 8–4

A SAP2000 model is shown in Fig. 8.67

Fig. 8.67 SAP2000 model
of Project 8–4

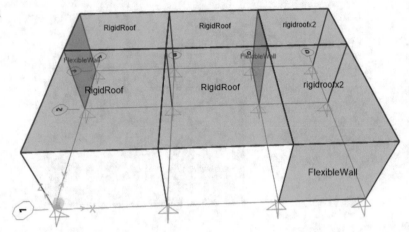

Project 8–5 Review the previous material in this chapter to fully understand the contribution to deflection from flexural behavior and from shear behavior. For a shear wall that is free on the top, as most of our traditional shear walls are, the following formula applies:

$$\Delta_{free\ on\ top} = \frac{Force \cdot height^3}{3EI} + \frac{3 \cdot Force \cdot height}{A \cdot E} \tag{8.30}$$

You are to study three different shear walls. Each wall has the following unchanging parameters:

- Thickness is 3in
- E modulus of elasticity is 3,604,997 lb./in^2
- G is the default SAP value

Then study three different walls:

- Width1 = 3 ft. Height1 = 12 ft. subjected to 10,000 lb. horizontal force along top
- Width2 = 12 ft. Height2 = 12 ft. subjected to 100,000 lb. horizontal force along top
- Width3 = 20 ft. Height3 = 12 ft. subjected to 300,000 lb. horizontal force along top

Figure 8.68 shows the three deformed walls. One is slender, one is square, one is squat. Fix the base in SAP.

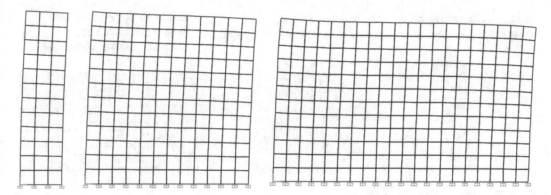

Fig. 8.68 Three walls of Project 8–5

   Calculate by hand the total deformation of each wall. Separate the shear deformation from the flexural deformation. Add the two for the total deformation. Comment on the magnitude of each deformation component.

   Repeat the three analyses in SAP2000. Comment on the agreement of the final, total deformation in each case.

   Write a brief conclusion. Include your response to the visual output from SAP and any lessons learned about the role of flexural deformation and shear deformation as the aspect ration (height/width) changes. Include a statement about why the given horizontal forces drastically changed between walls 1 and 3.

   Create a brief pdf of your report, with images from SAP and a succinct summarizing table.

   Figures 8.69 and 8.70 show typical student work for this Project 8–5.

Average Lateral Displacement is normalized to 141   Average Lateral Displacement is normalized to 3.3   Average Lateral Displacement is normalized to 1

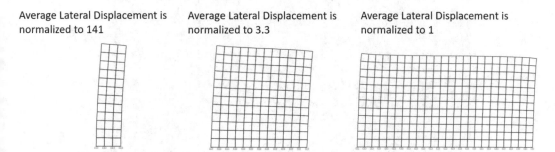

Fig. 8.69 Student work for Project 8–5

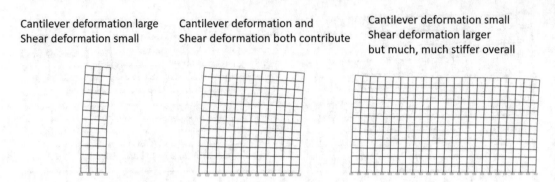

Cantilever deformation large   Cantilever deformation and   Cantilever deformation small
Shear deformation small   Shear deformation both contribute Shear deformation larger
                                                                      but much, much stiffer overall

Fig. 8.70 Another student work for Project 8–5

Project 8–6 An L-Shaped rigid diaphragm is shown in plan view in Fig. 8.71, dimensions are feet. Stiffness of each LFRS is shown as 1000 lb./ft., for example KSouth = 8000 lb./ft. Mass is uniformly distributed.

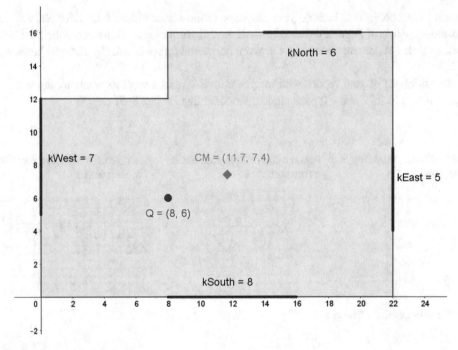

Fig. 8.71 Geometry of Project 8–6

a. Find the Center of Rigidity (CR)
b. If a Seismic Load is hitting the building in a South direction (flowing from kNorth towards kSouth), and if you were analyzing kWest, locate the CMeccentric based on the Code.
c. If Point Q = (8,6) were the CR and if no additional code eccentricity was used and a Seismic Load of 100,000 lb. is in South direction (flowing from kNorth towards kSouth) and it passed through the given CM, not the CMeccentric, what is the final force in the West LFRS?
d. If the DOF are at Point Q, find the third column of the Stiffness Matrix K by hand, not by SAP2000.

The solution to Project 8–6 part a is shown in Fig. 8.72.

Fig. 8.72 Solution to
Project 8–6 part a

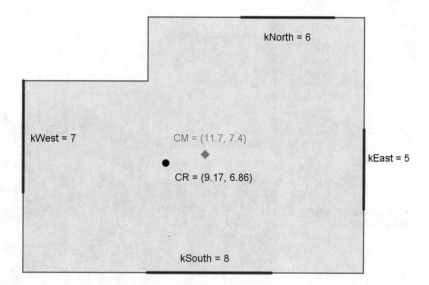

South load induces North force in KWest for Part1 of GPM. South load spins roof clockwise about CR, this induces South Force in Kwest. Lessen this reduction. Thus lessen the moment, thus wiggle CM to left, i.e. closer to CR. The solution to Project 8–6 part b is shown in Fig. 8.73.

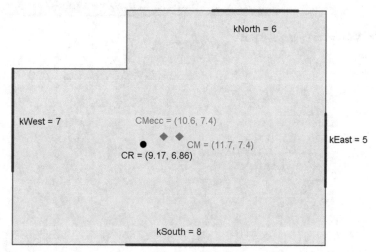

Use full 22 ft

Fig. 8.73 Solution to Project 8–6 part b

The solution to Project 8–6 part c is shown in Fig. 8.74.

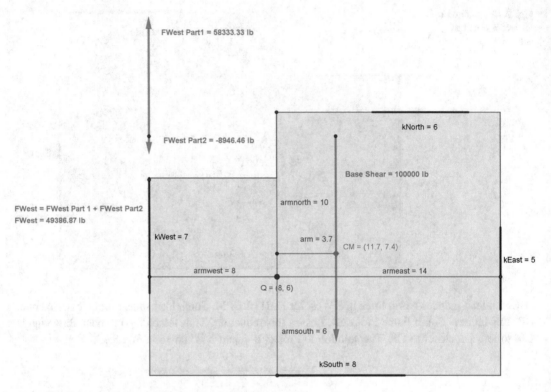

Fig. 8.74 Solution to Project 8–6 part c

The solution to Project 8–6 part d is shown in Fig. 8.75.

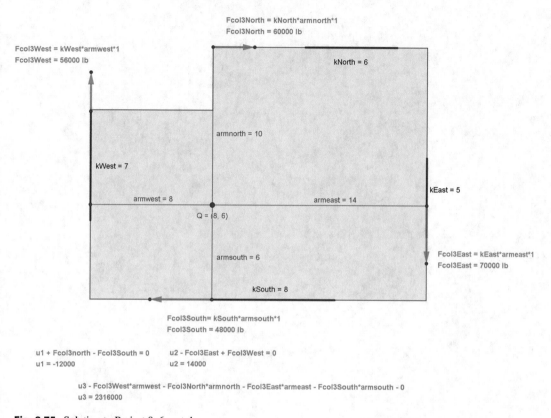

Fig. 8.75 Solution to Project 8–6 part d

Load Flow in Significant Buildings

9

In the third year architecture studio, three modules of a large project have been given to students. Module 1 was called "Replication". It was a guided precedent study of selected buildings, but the students were asked to specifically explain the gravity and the lateral load flow in the precedent building. They did this with models, figures and to a lesser extent with computer analyses.

For the architecture students, this was the first part of a studio course. In the latter part of the studio, the same students were asked to structurally model and explain their own personal designs. Yet, by starting with a known, significant structure, students were better able to respond to the difficult challenge of articulating structural behavior. For the architectural engineering students, this project was an in-depth study of many buildings which the students were unfamiliar with, thus it provided them a platform to discuss and to critique architecture from a structural engineering point of view.

For the instructors of the course, the task is surprisingly daunting. One reason is that for some of the buildings, the load path is not clear and unambiguous. It is helpful if published resources exist which explain the structural logic of significant buildings. Professor Guy Nordenson's book "Reading Structures" recently published by Lars Muller Press, is an ideal guide for such studies. It was used heavily in some of the precedent studies.

Another challenge facing instructors is the need to delicately, but constructively, critique and guide the students, as they attempt to grapple with structural logic. Students do indeed develop an intuitive structural sense, but that intuition must be nurtured and prompted by mentors.

Students tend to want to represent structure as an architectural model, i.e. they feel comfortable with showing a "image" of the building. They are much less comfortable with establishing a structural hierarchy, namely identifying which elements support subsequent portions of the building. But the benefits of this precedent study are tangible and lasting. Students plunge into deep investigations of a building they are interested in. They probe, ask questions, then refine their initial attempts. For the architecture students, all of these efforts are handsomely rewarded when it comes time to create their own structural schemes for their studio projects. They have seen what others have done, they feel more confident, and they can certainly use the language of the structural engineer when explaining load flow.

Module 2 was called "Innovation". This was a combination of several features of precedent buildings from Module 1, or a new refinement of an existing feature. In any case, the students were required to incorporate structural force resisting systems into their own architecture studio project. They were prompted to have the sizes of elements be reasonably to scale, although no algebra was used to size any particular member. They were asked to identify their material choices plainly and unambiguously. Most importantly, they had to explain clearly what was their Primary Structural

E. Saliklis, *Structures: A Studio Approach*, https://doi.org/10.1007/978-3-030-33153-5_9

System. Within the Primary Structural System, they had to clearly differentiate the *Gravity System* from the *Lateral Load System*.

Module 3 prompted the students to incorporate a tertiary structural element or system into their project. This tertiary structure ties the cladding back to either the primary or to the secondary structural system. Here, it was not important to quibble about what is primary and what is secondary. Nor was it important to have extremely realistic, practical connections. Rather, the students needed to imagine how might cladding be really attached to a building. If they could imagine a somewhat fanciful system, it could then always be improved by knowledgeable and experienced designers and builders. But if they cannot imagine such as system, perhaps it cannot be built at all! This argument was first presented to the author by William (Bill) Baker, of Skidmore Owings and Merrill fame. He also has discussed, with our students at Cal Poly, the hierarchy of primary versus secondary structure.

The prompt for Module 1 is presented here, as well as sample, representative student work. Many features of the modeling requirements in this prompt, and many of the significant buildings on the list, were suggested by Professor Catherine Wetzel. As the prompt suggests, the project is presented in two parts, part one is an initial attempt that is publically critiqued. Any glaring errors are corrected in the second part, and further refinement of the model is required for the second public critique. Physical models are required. All images are compiled into a single pdf which is used, alongside the physical models, for the public critiques.

Project 9–1: Replication This exercise is an exciting exploration that is qualitative, quantitative and fun. It is meant to encourage:

- The development of a technical literacy
- The promotion of an intuitive understanding of the flow of forces in a building
- The preparation of preliminary load flow models

The purpose of this project is to free the analyst from the challenges of designing a unique and new structural system, instead it asks for a description of an existing structural system. The challenge is to choose a building from the provided list, a building that elicits fascination, or that resonates somehow with one's own aesthetic or design ideas.

Build a model that captures the essential flow of gravity forces only, not lateral forces. This is not the exact architectural form of the building, rather it is the basic gravity skeletal system. The first model will be rough, coarse, essentially a 3D sketch. Use string, dental floss, hot glue, dowels, cardboard, duct tape. The second model in each module is more lasting but is not meant to be sophisticated, flashy or stylish. Here, springs, wood, cloth, screws and chipboard can be used. Laser cutting is fine. Pinned connections must be visibly rotating. Fixed connections must actually show deformations if loads are large, for example, the base of a shear wall would start to lift off and a spring restrains its motion, when the loads are large. Fixed connections can be created by adding cables at some distance from the connection, which induces a moment, note this is exactly the spirit of capturing the flow of force, not the specific form of the item being modeled. Ideally shear walls should deform in flexure and shear, but this might be completely impractical and is not required. Ideally, trusses would have no stiffness out of their own plane, again this is impractical, but think about it. Three dimensional pins could be modeled by two hinges, 90° apart from each other. Once again, this connection captures forces, not the form at all.

The second model in each module must have large deflections when you load it with your hands. Visible deformations are vital to the success of this model. The second, refined model is meant to add more detail, for example distinguishing between the primary and secondary structural system. It is meant to improve on any misconceptions presented in the first model. It is slightly more sophisticated, perhaps chip board, but it is not flashy, it is not meant for display. It is a descriptive tool. It should be large enough so that we see details from two meters away, but small enough that you can carry it to class.

When creating a document of the photos of your model, add arrows that pull away from each other to show tension and arrows that push together to designate compression. These arrows will show how load flows downward through the gravity system. Highlight what is in tension, what is in compression, what is in bending and which way it is bending. Scan all sketches and put them in a document with Figure Titles numbered and titled and placed below each figure.

Submit a PDF with images of the model and some text explaining the physics of gravity load flow. There must be several telling images of the original building in this PDF, images that describe the main structural features. Figures are numbered and titled uniquely, placed below the figure.

Then after the first critique, provide a refined, corrected model if there were errors originally. Submit a PDF with images of the model and clear unambiguous text explaining the gravity load flow. Please include distinctions between Primary and Secondary gravity systems. Again, there must be several telling images of the original building in this PDF, images that describe the structural features being modeled.

The List The list has grown and evolved over the several iterations that this course has been taught. Some have claimed that the list is the most interesting part of the course! Table 9.1 is the current list.

Table 9.1 List of Buildings for Precedent Study

Peckham Library	Alsop
Japan Pavilion Expo 2000	Ban
Centre Pompidou Metz	Ban
Nine Bridges Country Club Clubhouse	Ban
Centennial Hall	Berg
Danish Maritime Museum	BIG
Phoenix Central Library	Bruder
Wohlen School Auditorium	Calatrava
Wohlen School Entry Pavilion	Calatrava
Campo Volantin Footbridge	Calatrava
Train Station Liege	Calatrava
Bridge of Strings	Calatrava
Naples Railroad Station	Castiglioni
Temple as Space Vessel (yess!!)	Castiglioni
Porsche Museum	Delugan Meissl
Obes Warehouse	Dieste
Seagull Gas Station	Dieste
Institute of Contemporary Art	Diller Scofidio + Renfro
Eiffel Tower	Eiffel, Koechlin
Royal Victoria Dock Bridge	Eyre
Gateshead Millenium Bridge	Eyre
Footbridge Simone de Beauvoir	Feichtinger

(continued)

Table 9.1 (continued)

Jorba Laboratories	Fisac
Yokohama Terminal	FOA
Millenium Bridge	Foster
Hangar at Orly	Freysinnet
Suransuns Footbridge	Gartmann
Traversina Footbridge	Gartmann
Bridge over the Traversinertobel	Gartmann
Waterloo Terminal London UK	Grimshaw
Eden Project	Grimshaw
~~Farnsworth House~~	~~Goldsmith, Van der Rohe~~
Spiral Staircases St Thomas Chapel Columbia Univ, NY	Guastavino
Library and Learning Centre	Hadid
London Aquatic Centre	Hadid
Research Laboratory Vanafro	Happold
Hanover Pavilion	Herzog
1111 Lincoln Road Parking Structure	Herzog & deMeuron
College of Architecture, Univ of Minnesota	Holl
Kiasma Museum of Contemporary Art	Holl
MIT Simmons Hall Residence	Holl
Nelson-Atkins Museum of Art	Holl
School of Art Univ of Iowa	Holl
Sendai Mediateque	Ito
Izumo Dome	Kajima
Biblioteca Vasconcelos	Kalach
Yoyogi Gymnasium I Tokyo Olympics	Kawaguchi
Yoyogi Gymnasium II Tokyo Olympics	Kawaguchi
Luetschenbach School, Zurich	Kerez
Exeter Library	Khan
MO Museum Vilnius	Libeskind
Chiasso Shed	Maillart
Salginatobel Bridge	Maillart
Metropol Parasol, Seville	Mayer
Delft TU Library	Mecanoo
Jubilee Church, Rome Italy	Meier
Sony Center Berlin	Murphy, Jahn
Basento Bridge	Musmeci
Unesco Headquarters Auditorium	Nervi
Exposition Hall Turin	Nervi
Little Sports Palace Rome Olympics	Nervi
Dorton Arena Raleigh NC	Nowicki
Seattle Central Library	OMA
Munich Olympic Stadium	Otto
Multihalle Mannheim	Otto
Kansai Int'l Airport Arrival/Departure Hall	Piano
Padre Pio Church	Piano
Marie Tjibaou Cultural Center	Piano
Santa Fe Opera House	Polshek
Pompidou Center	Rice
Dulles Airport	Saarinen
Yale Hockey Rink	Saarinen

(continued)

Table 9.1 (continued)

ArtScience Museum	Safdie
Teshima Art Museum	Sasaki and Nishizawa
Toledo Museum of Art Glass Pavilion	Sejima, Nishizawa
New Museum of Contemporary Art, NY	Sejima, Nishizawa
Bochum Footbridge	Schlaich
Berlin Train Station	Schlaich
Lattice Communication Towers	Shukhov
Novgorod Exhibition Hall	Shukhov
Corning Glass Center 2000	Smith-Miller, Hawkinson
Hunt Library	Snohetta
Oslo Opera House	Snohetta
Broadgate Exchange House	SOM
Cadet Chapel US Air Force Academy	SOM
Millennium Park Footbridge	SOM
Richmond Oval	Structure Craft
Hiroshima Children's Libray	Tange
Osaka 1970 World Fair	Tange
Monument for the Third International	Tatlin
Hershey Ice Arena	Tedesko
Madrid Hippodrome	Torroja
Algeciras Market Hall	Torroja
Tokyo International Forum	Vignoly
Johnson Wax	Wright
Naiju Community Center	Yoh

Figure 9.1 is a study of the Yokohama Terminal building by FOA. The students began by focusing on trusses, which are not visually prominent, but do appear in online images of the structural system.

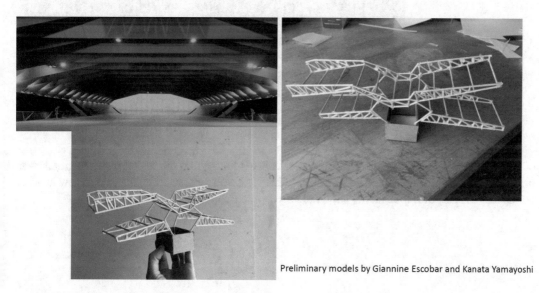

Preliminary models by Giannine Escobar and Kanata Yamayoshi

Fig. 9.1 Initial study of the Yokohama Terminal building

The students also began a stylistic representation of the main folded plate system. This is shown in Fig. 9.2.

Fig. 9.2 Study of folded plate system

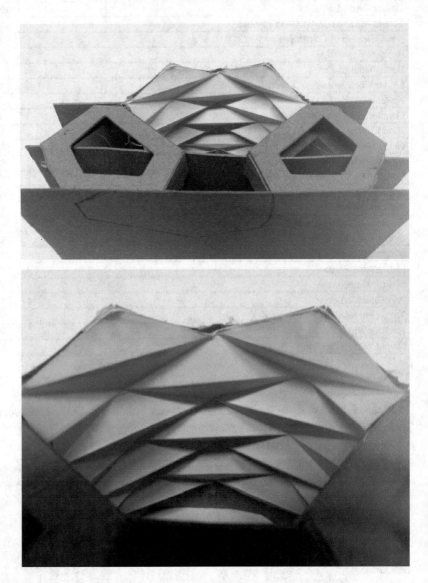

Their second iteration was much improved, because it focused on the folded plates themselves as primary. The second modeling effort was more refined and a bit more research uncovered new insights by the students. Figures 9.3 and 9.4 shows the second iteration of their study.

Fig. 9.3 Refined study of
the Yokohama Terminal
building

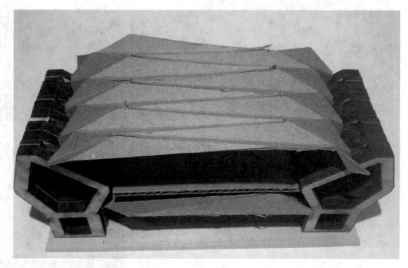

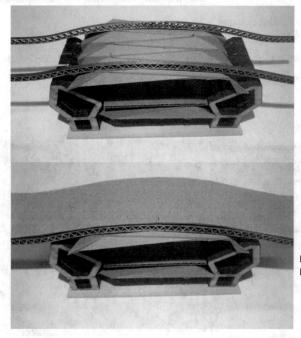

Final models by Giannine
Escobar and Kanata Yamayoshi

Fig. 9.4 Folded plate system with trusses

Figure 9.5 is an initial attempt by students to model the load flow of the Peckham Library by Alsop. Of course, the students began with internet-found images, and they attempted to model both the large cantilever subjected to gravity load, and the chevron bracing used to resist lateral loads.

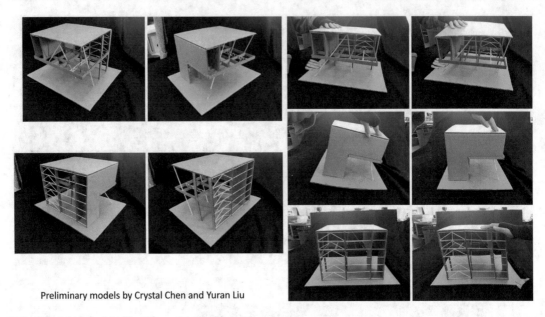

Preliminary models by Crystal Chen and Yuran Liu

Fig. 9.5 Initial model of load flow of the Peckham Library

Their second attempt focused more on the main cantilever, and less so on the chevron bracing. More detail is added to the second model, but it is clear this model is a structural one, not an architectural depiction Fig. 9.6.

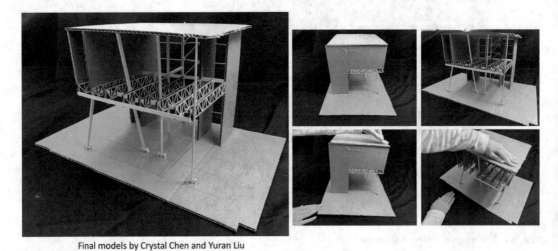

Final models by Crystal Chen and Yuran Liu

Fig. 9.6 Refined study of load flow of the Peckham Library

The Biblioteca Vasconcelos in Mexico City, by Kalach was chosen by two students because of the intriguing hanging book stacks. Students were able to find some images of construction drawings online, and they immediately focused on the massive beam/girder above the two-story book stacks. This is shown in Fig. 9.7

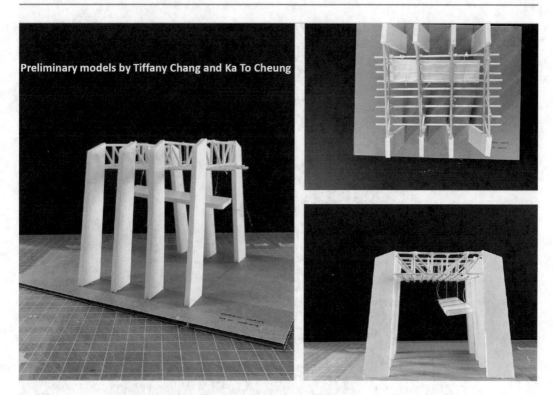

Preliminary models by Tiffany Chang and Ka To Cheung

Fig. 9.7 Initial model of the Biblioteca Vasconcelos

Their subsequent model was completely redone in Fig. 9.8. Notice the correct use of the beam/truss. It is technically not a Vierendeel as they first assumed, it is closer to the indeterminate T/C couple beam/truss studied in Chap. 6.

Fig. 9.8 Refined study of
the Biblioteca Vasconcelos

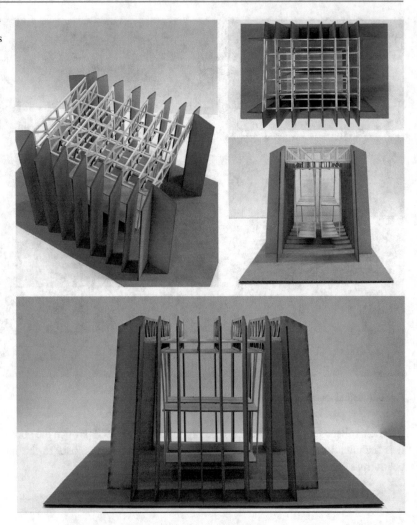

Figure 9.9 shows a preliminary study of the Porsche Museum by Delugan Meissl. In the students'
preliminary attempt, the large space truss was well modeled, but the huge slanted columns were not
fully structurally integrated yet.

Porsche Museum by Delugan Meissl

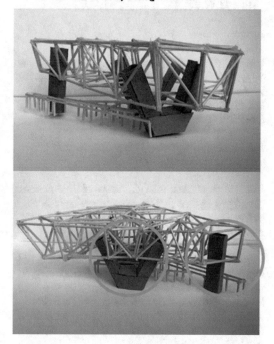

Models by Sydney Nguyen and Natalie Wanjek

The basement structure is supported by a grid system that is justified to two angles of the overall polygon shape.

Fig. 9.9 Initial model of the Porsche Museum

Their second iteration, shown in Fig. 9.10 is much more rational. The columns are beautifully integrated into the superstructure. The hierarchy of primary, versus secondary and tertiary structure has been studied.

STRUCTURAL ANALYSIS: MODULE 1, TRIAL 2

The primary structure consists of site cast concrete slabs in both the base structure and the superstructure above, consisting of a steel lattice and supported by 3 concrete pylons (two that split).

The base structure has coated precast columns and load bearing walls.

The structural cores or columns are reinforced, prestressed, self densifying concrete that fills in the pores to strengthen the concrete and make it self-polishing.

The superstructure's steel lattice resists lateral loads and covers the entire exterior depth as well as smaller interior depths for floor-to-floor levels. It's sitecast concrete slabs are constructed with holorib-plates (corrugated steel decks - aka trapezium plates) with insulation and waterproofing sheet.

The tertiary system includes the exposed self-densifying concrete, white-coated aluminum on the exterior of the superstructure, LSG (laminated safety glass) and structural glazing made with pillar bolt construction, and polished stainless steel on the soffit surface.

The projects took 3 years of construction and included and extensive team with not only an architect but also a site management architect, structural engineer, building services engineer, building physicist, etc.

Fig. 9.10 Refined study of the Porsche Museum

The Pompidou Centre-Metz by Ban was structurally modeled in Fig. 9.11 as a first attempt. The student initially focused on the important, and visually striking load flow into the unique columns.

POMPIDOU CENTRE - METZ
ARCE 316   TAMMY NGUYEN

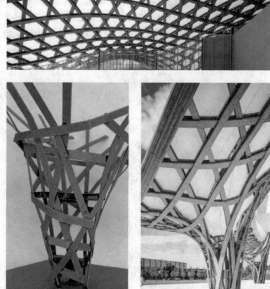

Glue Laminated Timber Roof
- Span up to 40 meters
- Self-supporting components

Gravity Load Transfer from Roof to Columns
- The roof comes down and turns into columns.
- The transition directs the load to meet the ground.
- Unique columns shapes and bracing resist Poisson Effects.

Fig. 9.11 Initial model of the Pompidou Centre-Metz

The complexity of the columns from a structural point of view was preliminarily modeled in Fig. 9.12.

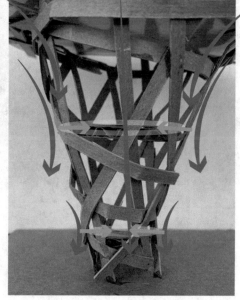

    Load Flow From Roof

⬭   Steel Brace

▬▬▬   Potential Deflection (Poisson Effect)

▬▬▬   Resisting Force

Fig. 9.12 Initial model of Pompidou Metz column

The second iteration attempted to capture the most intriguing element of the structure, namely the funicular shape of the roof. What would have been a hanging net in tension became (when inverted), a slender, compression-only grid shell. Without critiques from the mentor and from peers, the student would not have been able to understand the funicular shape. The funicular shape was studied qualitatively in Fig. 9.13. The load flow into the columns uses the correct doubled arrows in Fig. 9.14.

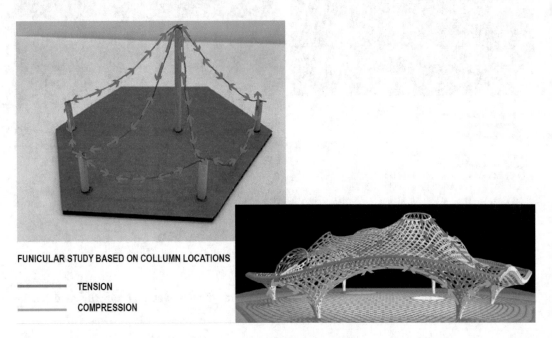

Fig. 9.13 Further study of Pompidou Metz roof

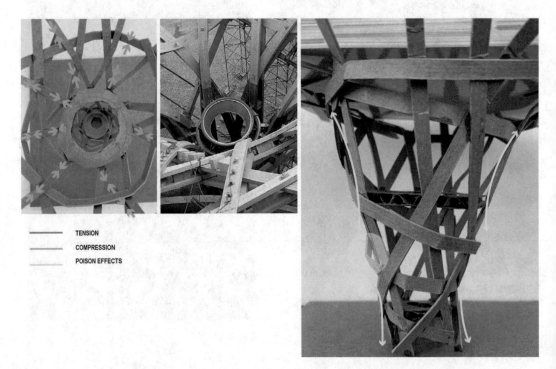

Fig. 9.14 Further study of Pompidou Metz column

The Hunt Library by Snohetta was studied in Fig. 9.15. This student's first effort was remarkably detailed. Only subtle improvement were made for the second iteration.

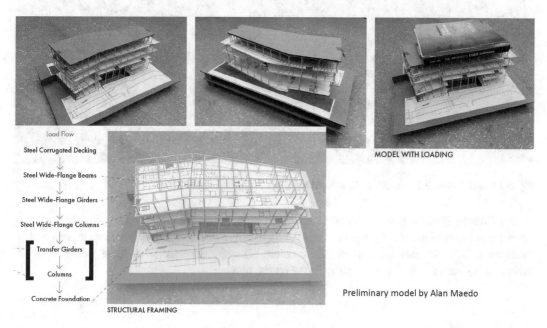

Fig. 9.15 Initial study model of the Hunt Library

In this first set of models, a clever addition was the use of a printed floor plan on the base of the model. It adds much to the story and it clearly states that this is a structural, not an architectural model. These plans are very clearly shown in Fig. 9.16.

Fig. 9.16 Refined study model of the hunt library

The second iteration here was not substantially different than the first. The understanding of the lateral supports was deeper in Fig. 9.17, with an unambiguous load path from the diaphragms to the shear walls. Also, bracing of the columns was suggested in Fig. 9.17

Fig. 9.17   Shear walls and columns of the hunt library

The Chiasso Shed by Robert Maillart was studied in Fig. 9.18. One major reason for choosing this structure is because the late Professor David Billington had developed very interesting pedagogical modules to help students understand this structure. The student was given these materials as an invaluable resource, which is not readily found on the internet.

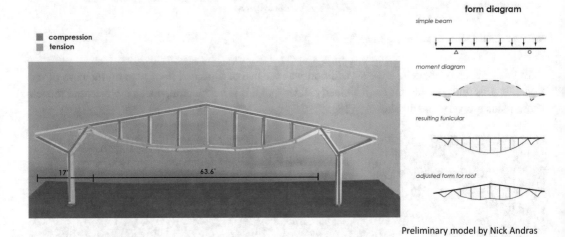

Preliminary model by Nick Andras

Fig. 9.18   Initial study model of the Chiasso Shed

Although the preliminary model in Fig. 9.18 was reasonable, the student was challenged to look deeper at the bending moment diagram as a form generator. Also, the student was asked to consider out-of-plane stability of the primary arch/trusses. The second iteration is shown in Fig. 9.19 which shows a beautiful and functional structural model.

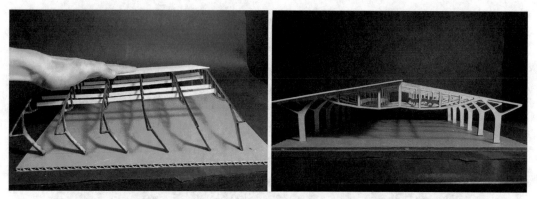

Final model by Nick Andras

Fig. 9.19 Further study of the Chiasso Shed

Figure 9.19 is a good transition point to Module 2 of the studio. Here, the students were asked to use some of the ideas seen in the precedent studies of Module 1, either their own or their peers' work. Then, they were prompted to incorporate some of these structural insights into a part of their own studio work.

For some, this was extremely liberating because they understood the physics of the precedent studies, and they knew well their own needs in their studio model. The student who studied Maillart's Chiasso Shed was able to integrate Maillart's ideas into his own project. Even though the studio project is decidedly abstract, sculptural and mostly theoretical, structural rationalism is still necessary. Figure 9.20 shows the Maillart idea and Fig. 9.21 shows it within a larger part of the studio project.

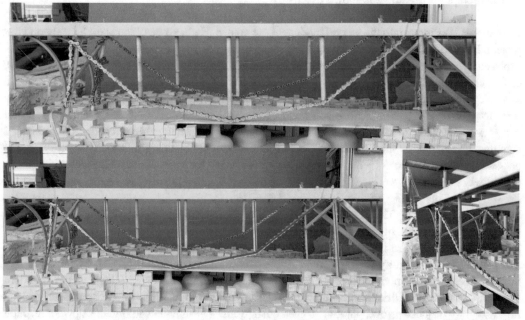

Incorporating R. Maillart into studio project: by Nick Andras

Fig. 9.20 Closeup of Chiasso Shed system incorporated into studio project

Fig. 9.21 Overview of Chiasso Shed system incorporated into studio project

The prompt for Module 2 is shown here:

Project 9–2 We move into the second module of our course. Here, you will focus exclusively on your project, but you will be asked to draw from the rich examples you have seen in the precedent works explored by you and by your peers. Think of the interesting cantilevers, walls, tapered beams, shells, cables, clever moment frames (Myron!), towers, all of the fun primary and secondary systems we have seen. Now make them your own! You can use some of these terrific ideas directly, but you must apply them to your own structure. Before you plunge in though, remember some of the curious and odd things we saw, like no clear lateral force resisting system, questionable decisions about slanted columns, discontinuous shear walls, irrational forms, deceiving cladding that looks structural but is not, strange forms that do not reflect physical load flow. Avoid those and focus on the wonderful things we did see.

Module II, Innovation This is to be a combination of several existing features of precedent buildings (yours or your peers'), or a new refinement of an existing feature. Whatever you do, you must incorporate this structure into your own architecture studio project. Have the sizes of elements be reasonably to scale. Identify your material choices plainly and unambiguously. Clearly show Primary Structural System for Gravity and propose a hierarchical ordering of Primary Gravity, and Secondary Gravity. Of course, this is debatable, but it helps clarify your thinking. Show primary compressive load flow as tightly matched arrows pointing at each other. Show a clear and unambiguous upward reaction from the ground. If you have tension elements helping you, (hooray!) show that path with tightly matched arrows that pull away from each other, and show us where this terminates. Completely define the gravity system for all the major parts of your building, the big picture, not every minute detail. Again, this system must have a primary system and a secondary system.

Then, on your project establish only a Primary Lateral Force Resisting System (LFRS) in each of the two major perpendicular directions of your building. Most often this is either shear walls, braced frames (trusses) or moment frames. But shells coming down to the ground are very good for LFRS. Elevator shafts, mechanical shafts, stair wells are also terrific as LFRS. But a single cantilever column is absolutely not acceptable as an LFRS. Do not create any Secondary LFRS, but you must describe what is happening in each of the two primary directions.

Upload a pdf that presents your name, a few descriptive images of your building so far, and a few initial questions that you may have about structure. For example, how might you establish a column grid for gravity? How might a long cantilever be supported? How might curved surfaces such as a shell support themselves? How might parts of a building hang off of primary gravity systems? How might you perforate a wall to let light in? How might large three hinged arches help? How might cables help? Initially, keep all the information about your building big picture, and focus on the most burning questions about gravity systems. Don't worry too much about lateral unless you feel very comfortable with your gravity system. Don't worry about cladding right now. Now is the time for big shifts if needed, not fine tuned small changes. Convince yourself it will stand up due to gravity and due to lateral loads.

The third year architecture students focus on wall sections in their Architectural Practice class. Also, the architecture department asked the architectural engineering department for help in better understanding how cladding is attached to a building. Thus, Module 3 of this project came about, and the prompt for this is presented.

Project 9–3

Module III, Tertiary Structural System and Cladding This will tie in beautifully to your practice class. Consider what specific cladding is applicable to a portion of your project. Not all the cladding, but work out one interesting part of the cladding problem. If louvers move, have them be flexible in your model. If glass is used, represent it with colored paper and distinguish it from mullions by having the mullions a different material.

You certainly may use an existing detail. This is not cheating, artists call this "appropriation". But, it is absolutely necessary that you make this detail applicable to your own structure.

Each of you must make a mockup model of your detail. It can be cardboard, duct tape, not elegant. But it must clearly and unambiguously show how the cladding is structurally connected back to the building. The structural elements are called tertiary elements in this course. They typically tie back to secondary structure in the main building. Tricky perforations must be addressed, for example if you have a metal screen as well as a glass skin, how would the metal screen penetrate the glass and connect back to the secondary structure? If you have a substantial gap between your cladding and your building, you must clearly and unambiguously show the cantilever elements that bridge this gap.

The following figures show student work for Project 9–2 (Module 2 primary and secondary Gravity and Lateral Force Resisting Systems) and from Project 9–3 (Module 3 Tertiary Structure and Cladding) from several iterations of this studio. Some of the structural elements are clearly gleaned from the precedent studies, others are re-invented in a somewhat fanciful manner. The point of these exercises is to convince the students and the jurors that a solid understanding of structural load flow exists, both for gravity loads and for lateral loads. Students learned by doing, which is the motto of our University.

Fig. 9.22 gives an overview of a third year student's project

*STRUCTURAL INTEGRATION -- **ANTS AND AGENTS***

Violeta Smart | ARCE 316

Fig. 9.22   Overview of a third year student's project

Even though the exterior of the building makes it seems very complex, a regular and orderly structural grid was established, which was liberating, not limiting, to the student. The Primary Structural System is shown in Fig. 9.23.

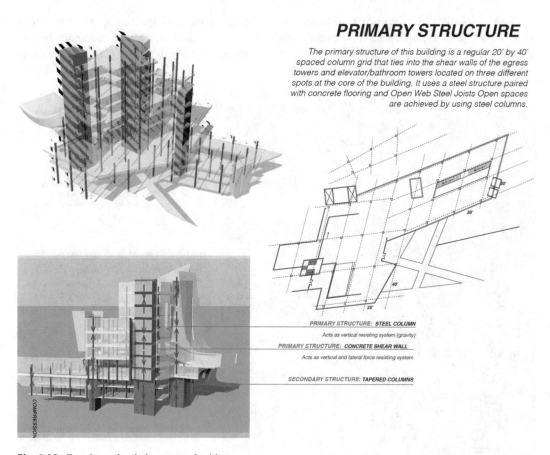

PRIMARY STRUCTURE

The primary structure of this building is a regular 20' by 40' spaced column grid that ties into the shear walls of the egress towers and elevator/bathroom towers located on three different spots at the core of the building. It uses a steel structure paired with concrete flooring and Open Web Steel Joists Open spaces are achieved by using steel columns.

PRIMARY STRUCTURE: **STEEL COLUMN**
Acts as vertical resisting system (gravity)

PRIMARY STRUCTURE: **CONCRETE SHEAR WALL**
Acts as vertical and lateral force resisting system

SECONDARY STRUCTURE: **TAPERED COLUMNS**

Fig. 9.23 Regular and orderly structural grid

Fig. 9.24 shows the physical models that were used to better understand the structural system, and to easily communicate challenges to the studio mentor.

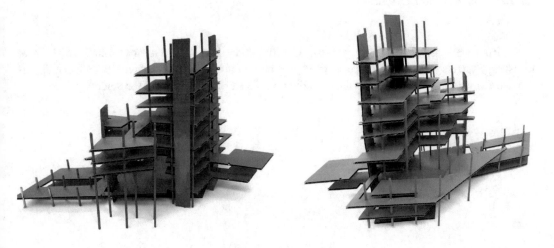

primary structure and lateral support physical model

Fig. 9.24 Physical models to further clarify structural grid

Figure 9.25 shows the Secondary Structural System. Engineering professors enjoy quibbling about what is truly primary, and what is secondary, such arguments should be avoided. Students need to establish hierarchical thinking in their approach to structure, what are the major elements? How do subsequent elements frame into the major ones?

SECONDARY STRUCTURE

The secondary structure of this building is the space truss that encloses the collaborative cloud, which exists as a volume inside the building (and slightly cantilevering out, supported by a tapered column)

The render above, shows how this space truss comes from the outside of the building, to the inside .

Fig. 9.25 Secondary structural system

Students were asked to clearly separate Gravity Force Resisting Systems from Lateral Force Resisting Systems (LFRS). In practice, these can be combined, but for pedagogical reasons it was decided to unambiguously separate the two. Figure 9.26 shows the LFRS of the student project.

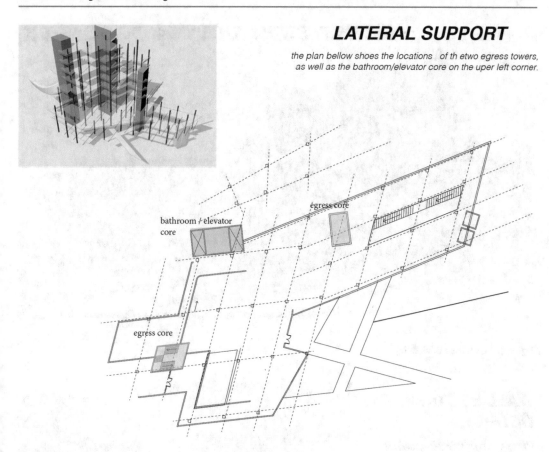

LATERAL SUPPORT

*the plan bellow shoes the locations of th etwo egress towers,
as well as the bathroom/elevator core on the uper left corner.*

Fig. 9.26 Lateral force resisting system of student project

Figure 9.27 gives an overview of the cladding and Fig. 9.28 gives some detail, via a wall section, of how this cladding might be attached back to the building using Tertiary Structure.

TERTIARY STRUCTURE: CLADDING SYSTEM

Fig. 9.27 Overview of cladding

WALL SECTION DETAIL

PERFORATED METAL MESH SKIN FOR ENVIRONMENTAL MITIGATION.

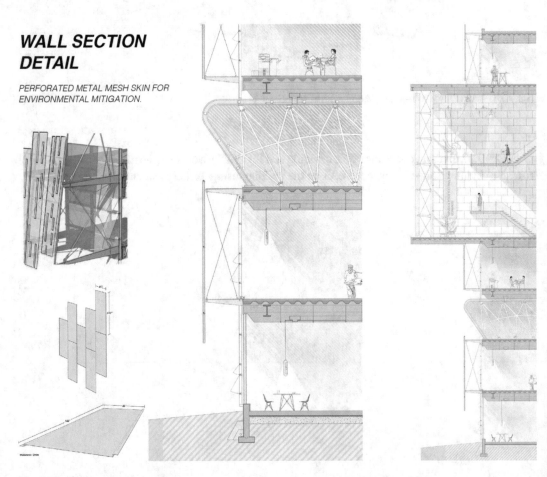

Fig. 9.28 Tertiary structural system

Figure 9.29 shows the student's final physical model and Fig. 9.30 shows the mockup of the cladding detail.

Fig. 9.29 Final physical studio model

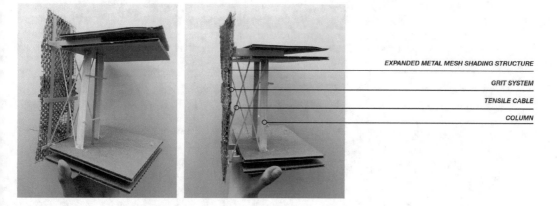

EXPANDED METAL MESH SHADING STRUCTURE

GRIT SYSTEM

TENSILE CABLE

COLUMN

Fig. 9.30 Student mockup of cladding detail

In yet another third year architecture studio, the three modules were again applied. The following figures show some representative work. Figure 9.31 clearly identifies Primary, Secondary and Tertiary structure in the students' studio project.

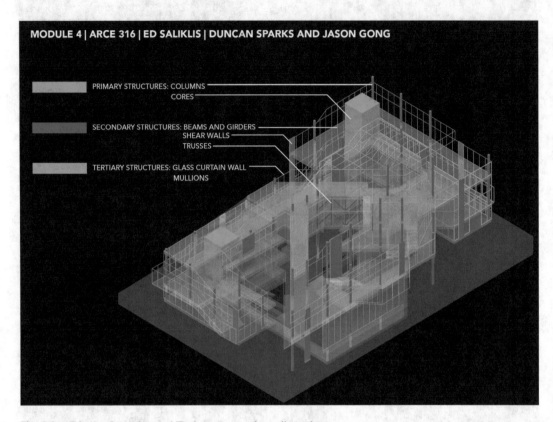

Fig. 9.31 Primary, Secondary and Tertiary structure in studio project

Many of the major details of Fig. 9.31 were incorporated into a physical model, which is shown in Figs. 9.32 and 9.33. One advantage from the students' point of view, for making such a model, is that during their final studio critique, they have very descriptive stories to tell the jurors, stories that describe how the structure was integrated into the architecture. These physical models clearly help to tell that story.

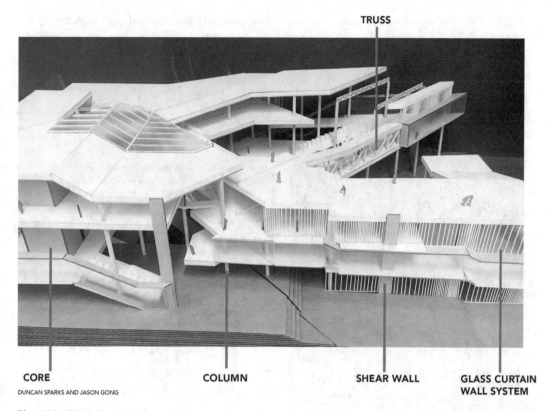

TRUSS

CORE COLUMN SHEAR WALL GLASS CURTAIN
WALL SYSTEM

DUNCAN SPARKS AND JASON GONG

Fig. 9.32 Physical model of structural systems

Fig. 9.33 Close-up view of physical model of structural systems

Some of the tertiary structural ideas were worked out, and literally "fleshed out" with rough, 3D mockups built by the students. Figure 9.34 shows one such model, with a corresponding wall section.

FLOOR SLAB AND DROPPED CEILING

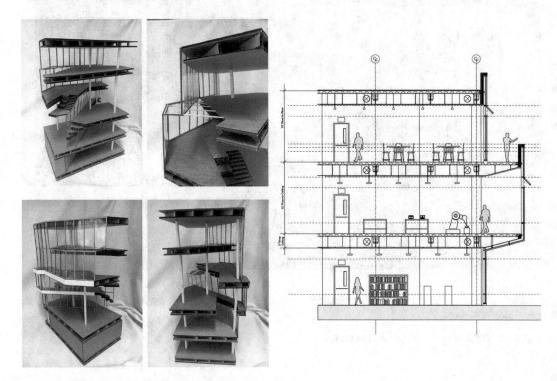

Fig. 9.34 Tertiary structure mockup and section drawing

After group critiques of the tertiary structure and the proposed cladding system, the students refined their digital model and developed extremely descriptive drawings of this Module 3 effort. Figure 9.35 shows one such drawing.

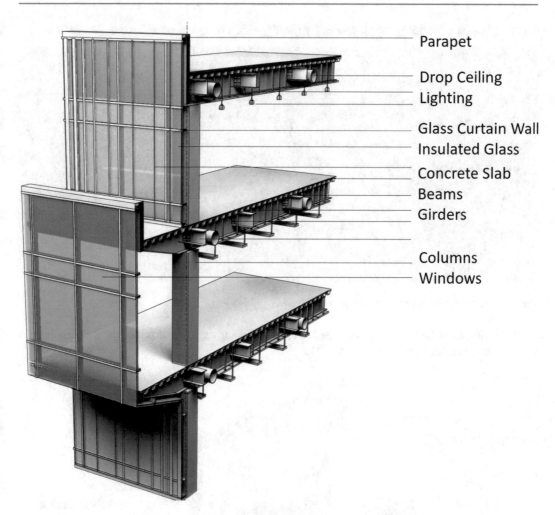

Parapet

Drop Ceiling
Lighting

Glass Curtain Wall
Insulated Glass
Concrete Slab
Beams
Girders

Columns
Windows

Fig. 9.35 Extremely refined drawing of tertiary structure and cladding

The prompt clearly asks for a 3D physical model of the tertiary structure and a portion of the cladding. This is uncharted territory for most students, and to be honest, for most faculty. Thus, imaginative efforts that seem buildable are much preferred over simply copying some known details from the internet. An example of a well thought out student detail model is shown in Fig. 9.36.

DETAIL FACADE CONNECTION | STICK BUILT MULLION WITH INSULATED GLASS UNIT AND FLOOR CONNECTIONS

Fig. 9.36 Refined mockup of tertiary structure and cladding

Final critiques of that physical model lead to an eye-popping exploded axonometric drawing created by the students, which is shown in Fig. 9.37.

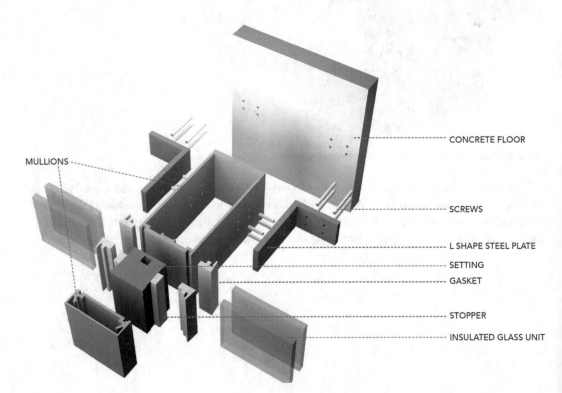

Fig. 9.37 Exploded axonometric drawing of tertiary structure and cladding

One of the most challenging studios that the author has had the privilege of working in, is a particularly *avant garde* studio that starts with found objects at a waste management landfill, and proceeds to create fanciful, sculptural architectural constructions. It is difficult to find means of convincing students to adhere to strict grids in such a studio, the most that is hoped for is moments of stability and rational load flow to the ground. Figure 9.38 shows the cover of one portfolio from this studio. This alone provides a sense of the structural challenges ahead!

RENDEZVOUS
JOHN LANGE'S FINAL PROJECT
MODULE IV
FINAL PRESENTATION
Alan Maedo | Lange Gange
ARCH 316-07 | Saliklis | Spring 2019

Fig. 9.38 Cover page of *avant garde* studio project

To help articulate some structural logic, it was suggested to superimpose section drawings with slightly grayed photographs of the complex model. One such superposition is shown in Fig. 9.39

Overpass: Structural Steel Framing
Gravity Loading System

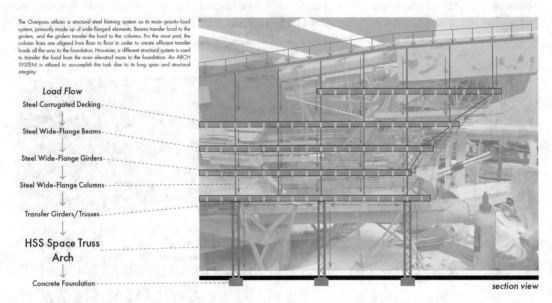

The Overpass utilizes a structural steel framing system as its main gravity-load system, primarily made up of wide-flanged elements. Beams transfer load to the girders, and the girders transfer the load to the columns. For the most part, the column lines are aligned from floor to floor in order to create efficient transfer loads all the way to the foundation. However, a different structural system is used to transfer the load from the main elevated mass to the foundation. An ARCH SYSTEM is utilized to accomplish this task due to its long span and structural integrity.

Load Flow
Steel Corrugated Decking
↓
Steel Wide-Flange Beams
↓
Steel Wide-Flange Girders
↓
Steel Wide-Flange Columns
↓
Transfer Girders/Trusses
↓
HSS Space Truss
Arch
↓
Concrete Foundation

section view

Fig. 9.39 Initial drawing of primary structural system

This technique of highlighting the Primary Structural System was extended to a remarkable set of drawings. Figure 9.40 highlights the Primary Gravity Force Resisting System and Fig. 9.41 shows the Primary Lateral Force Resisting System.

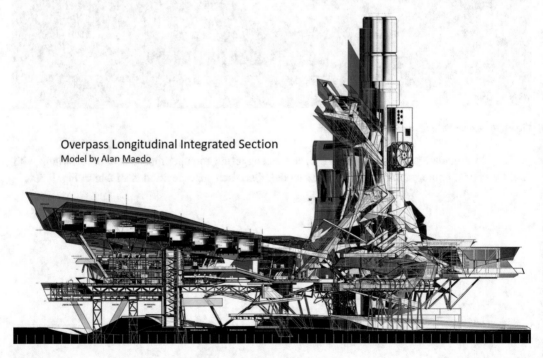

Overpass Longitudinal Integrated Section
Model by Alan Maedo

Fig. 9.40 Extremely refined drawing of primary structural system

Fig. 9.41 Close-up view of drawing of primary structural system

The Cladding System of this project is shown in an overview drawing in Fig. 9.42. Then, a detailed drawing of how the cladding connects to the building through the Tertiary System is shown in Fig. 9.43, with photos of the physical mockup in Fig. 9.44.

Fig. 9.42 Overview of cladding in complicated studio project

Cladding Materiality

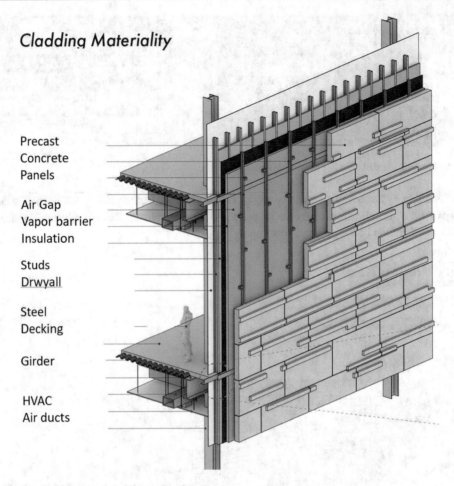

Precast
Concrete
Panels

Air Gap
Vapor barrier
Insulation

Studs
Drwyall

Steel
Decking

Girder

HVAC
Air ducts

Fig. 9.43 Detailed drawing of cladding and tertiary structure

Physical Studies
Model Photos

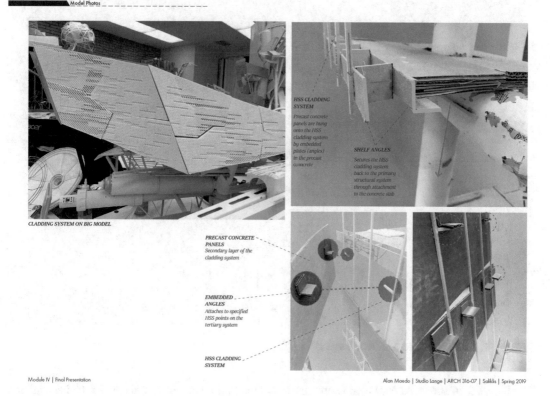

CLADDING SYSTEM ON BIG MODEL

HSS CLADDING SYSTEM
Precast concrete panels are hung onto the HSS cladding system by embedded plates (angles) in the precast cconcrete

SHELF ANGLES
Secures the HSS cladding system back to the primary structural system through attachment to the concrete slab

PRECAST CONCRETE PANELS
Secondary layer of the cladding system

EMBEDDED ANGLES
Attaches to specified HSS points on the tertiary system

HSS CLADDING SYSTEM

Module IV | Final Presentation

Alan Maedo | Studio Lange | ARCH 316-07 | Saliklis | Spring 2019

Fig. 9.44 Physical mockups of cladding and tertiary structure

Two interesting spatial structures are described next. The first is a grid shell that undulates over an extremely large expanse. These are the students that were inspired by the Yokohama Terminal building, studied in Fig. 9.4. Their studio project begins with Fig. 9.45 and it is clear that they borrowed some ideas, but were extremely independent and creative in their own work.

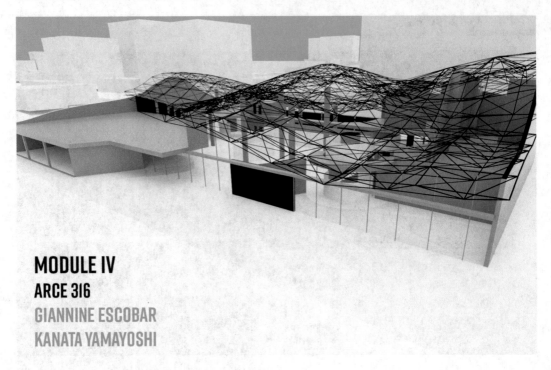

MODULE IV
ARCE 316
GIANNINE ESCOBAR
KANATA YAMAYOSHI

Fig. 9.45 Overview of gridshell studio project

Figure 9.46 shows an exploded axonometric of their entire structural concept. An ETFE cladding was envisioned to rest above the vast grid shell.

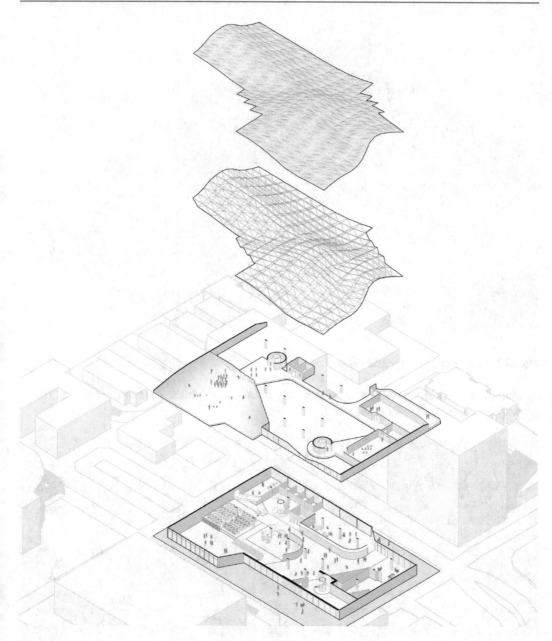

Fig. 9.46 Exploded axonometric drawing of gridshell studio project

It was not perfectly clear to the instructor whether or not the grid shell could act as a diaphragm that flows into a lateral force resisting system, but the students were allowed to assume this was feasible, due to the very high stiffness of the undulating surface. Thus, the primary lateral force resisting system shown in Fig. 9.47 has this roof flowing into massive shear walls.

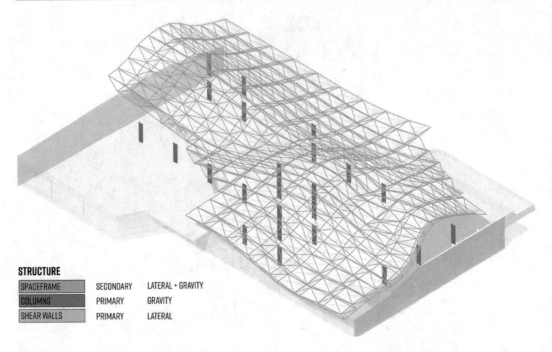

STRUCTURE

SPACEFRAME	SECONDARY	LATERAL + GRAVITY
COLUMNS	PRIMARY	GRAVITY
SHEAR WALLS	PRIMARY	LATERAL

Fig. 9.47 Primary structure of gridshell roof

Multiple smaller shear walls were strategically placed on the main floor of their space. This is shown with a physical model in Fig. 9.48.

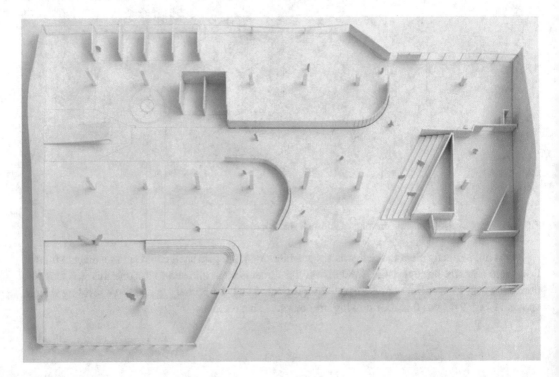

Fig. 9.48 Primary structure below gridshell roof

To further test their ideas, the students created a physical model of the undulating grid shell roof, which nicely rests on the main floor. This is shown in Fig. 9.49.

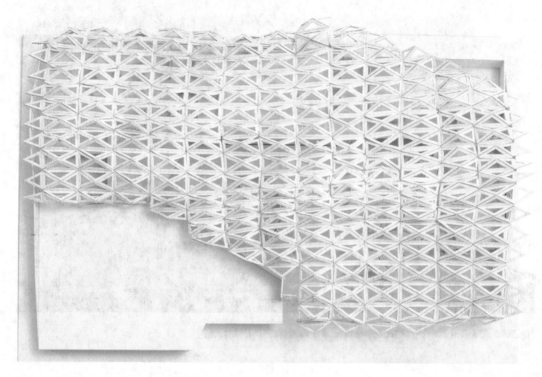

Fig. 9.49 Physical mockups of gridshell roof

Digital and physical models of the tertiary structure which will hold the ETFE panels were very helpful in this deep investigation of how to actually assemble such a complex form. These are shown in Fig. 9.50 as a digital study, in Fig. 9.51 as a hybrid digital/physical study, and in Fig. 9.52 as a physical mockup.

ETFE DETAIL INTEGRATION

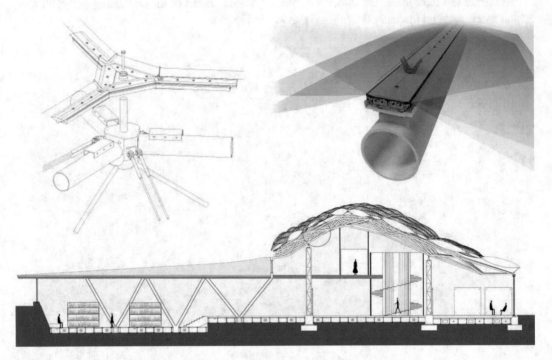

Fig. 9.50 Detailed drawing of tertiary structural system for ETFE cladding

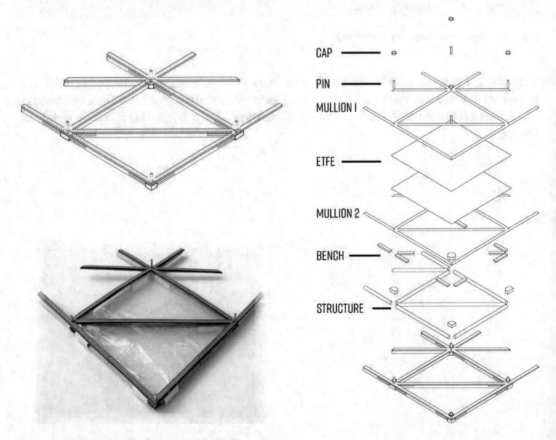

CAP

PIN

MULLION I

ETFE

MULLION 2

BENCH

STRUCTURE

Fig. 9.51 Hybrid digital/physical study for ETFE cladding

Fig. 9.52 Physical mockups of ETFE cladding

The student project in Fig. 9.53 combined grid shell behavior with monolithic reinforced concrete thin shell structure in a remarkably sophisticated manner.

Project by Yuran Liu

Fig. 9.53 Overview of monolithic shell and gridshell studio project

Figure 9.54 shows how the thin shell reinforced concrete roof folds over to blend into lower story walls. The Instructor tried many times to dissuade the student from making such sharp turns, she steadfastly refused to listen and maintained her design decision throughout the rest of the project. The suspicious turns of the roof are shown in Fig. 9.54. They are in contrast to an otherwise smoothly flowing, structurally rational and beautiful roof.

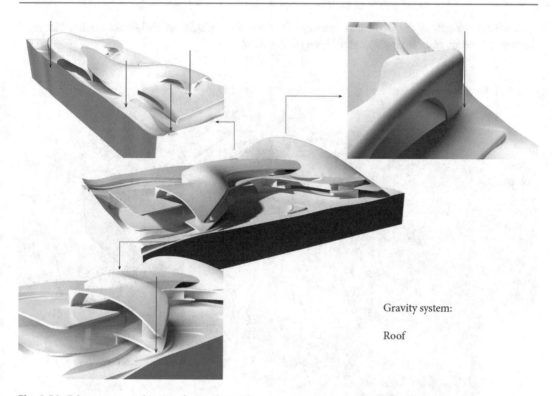

Gravity system:

Roof

Fig. 9.54 Primary structural system for gravity loads

One of the most striking features of this particular project is the extraordinary detail provided in the digital studies. Figure 9.55 shows how the student referred back to her precedent study of flat slab buildings and column capitals, in her deep dive study of the studio project.

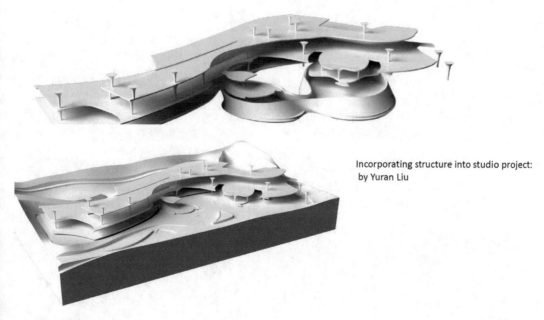

Incorporating structure into studio project:
by Yuran Liu

Fig. 9.55 Detailed study of precedent slabs and studio slab

A telling example of the virtuosic renderings is shown in Fig. 9.56, the Instructor couldn't know if this was a rendering, or a photograph of a physical model.

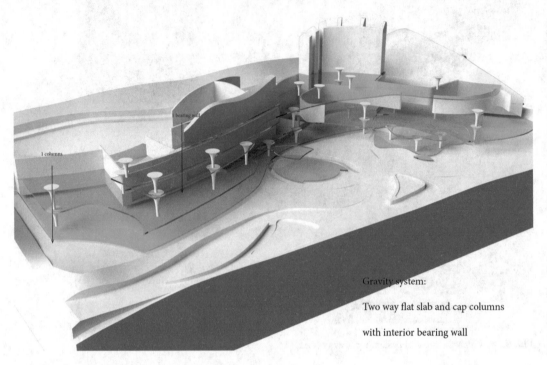

Fig. 9.56 Incorporation of bearing walls and columns

Figure 9.57 describes the lateral force resisting system. The undulating roof is laterally supported by shear walls. The Instructor attempted to get the student to use the egg-shaped grid shells of Fig. 9.57 as another lateral force resisting system, but the student again resisted advice and insisted that the grid shells be visually and physically separate from the roof.

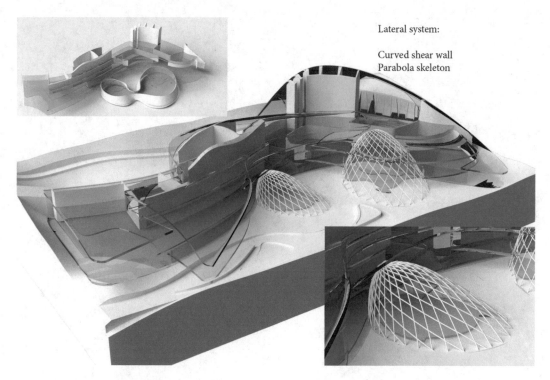

Lateral system:

Curved shear wall
Parabola skeleton

Fig. 9.57 Detailed study of lateral force resisting system

Frame Refinement

10

The qualitative and quantitative explorations of frames in this chapter owe a great deal to two German professors and educators, Heino Engel and Curt Siegel. Engel's landmark book *Tragsysteme* is well known to many because of its astonishing illustrations, which explore different types of structural forms, including frames. Engel qualitatively explored frame behavior in a visual, exploratory manner. Less well known is the work of Professor Curt Siegel of Stuttgart University, who made a major contribution to the study of structures, including frames and shells, with a book called *Structure and Form in Modern Architecture*. Siegel called his novel approach to teaching structures, which joined together static analysis of structures and design of structural forms, *"Tragwerklehre."* This can be translated as "teaching on structures" or the translation of statics and elastic behavior of solids into actual structural elements of a building. This chapter on frames is indebted to both of these great pedagogical works. Structurally rational frames have multiple aesthetic possibilities. Consider a portion of a concrete frame shown in Fig. 10.1, notice the placement of the nodes above the column along the top of the beam line. Notice Z is upwards in this model.

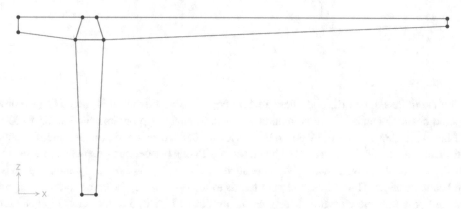

Fig. 10.1 Portion of a concrete frame

Replicating this as a mirror stitches the nodes together at what essentially is a hinge at the mid-span of the long beam. One bay in the XY plane is now created and is shown in Fig. 10.2, which has a short exterior cantilever, and a long interior span.

© Springer Nature Switzerland AG 2020
E. Saliklis, *Structures: A Studio Approach*, https://doi.org/10.1007/978-3-030-33153-5_10

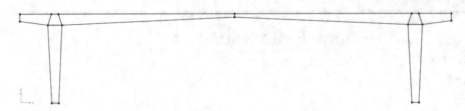

Fig. 10.2 Mirroring to create a bay with central hinge

Vertical replication will lock the base of the column nodes to the beam nodes above the column. Any number of replications can be made.

Fig. 10.3 Vertical replication to create a multi-story frame

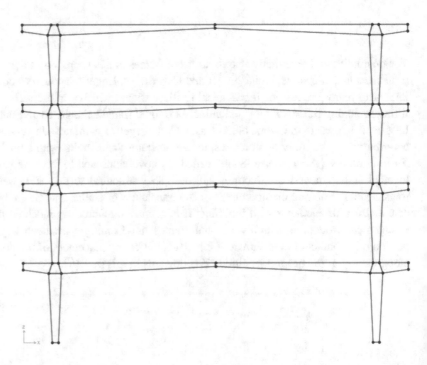

Within the finite element method, there are two choices for modeling this frame. Choice one is to use tapered beam elements. Many commercial structural analysis programs such as SAP2000 have tapered beam elements, they are made up of previously defined prismatic (non-tapered) beams, and they are created such that the new tapered beam starts with one pre-defined prismatic cross section and ends with another. However, some programs such as Karamba, do not easily lend themselves to tapered beam elements. Thus, choice number two is to model a tapered beam or column as a mesh of area or shell or membrane elements. Changing the outline of Fig. 10.3 to the series of shell elements shown in Fig. 10.4 is known as "meshing" an area or discretizing one large area into many smaller, but roughly uniform, finite elements. Mesh refinement is good practice, as the finite element method requires uniformly sized, not greatly distorted elements to approach theoretically correct answers. Meshing is somewhat of an art, but the following mesh shown in Fig. 10.4 is roughly what one seeks when refining areas. Many wonderful plugins such as Weaverbird can be used to create more sophisticated, uniform and finely grained meshes within the Rhino environment. Figures 11.15 and 11.15 of the following chapter and used Weaverbird to create beautiful, uniform meshes.

Fig. 10.4 Finite element
mesh of concrete frame

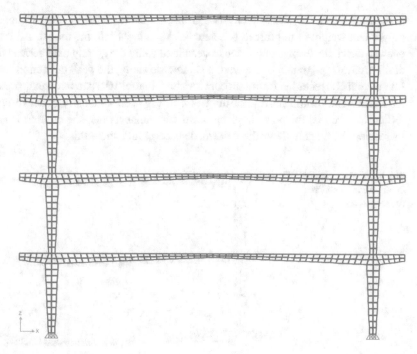

Figure 10.5 shows a thin, but possibly acceptable profile to the frame in a 3D view.

Fig. 10.5 3D view of
concrete frame

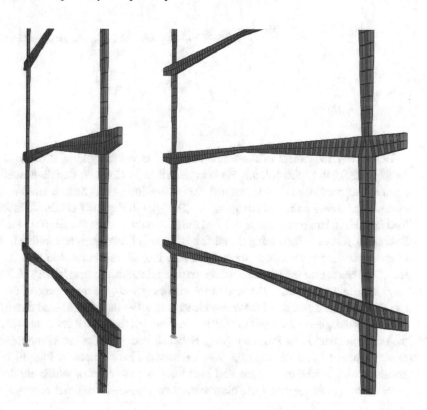

A Dead Load, 2D planar analysis can be run, if the frame is some reasonable thickness and the proper unit weight of reinforced concrete has been input into the model. After the dead load analysis, one can check the displacement due to dead load using the greatly exaggerated deformed shape feature of the analysis programs. If this were a parametric study, the peak deflection and overall weight of the frame would be recorded for future comparisons. Figure 10.6 shows an exaggerated deformed shape of the frame. The central portion of the beam in Fig. 10.6 is clearly concave up, but it is much more difficult to see that the portion of the beam that connects into the column is actually concave down. Increasing the magnitude of the deformed shape would show this.

Fig. 10.6 Exaggerated deformation of concrete frame

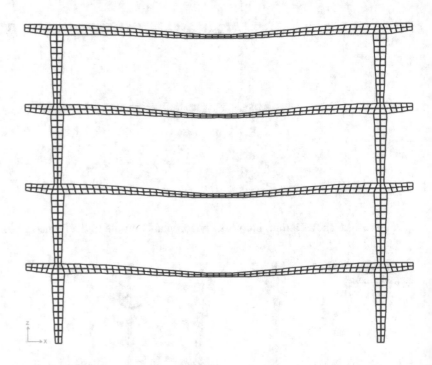

Before any refinement of the frame occurs, it is worth looking at the load flow from the beam, through the joint, to the column. As was described in Chap. 9, double-headed arrows best describe tension and compression. Arrows that pull away from each denote tension, and arrows that push towards each other represent compression. The joint in Fig. 10.7 exhibits a bit of "turbulence" i.e. the load flow is not laminar, because the load must literally turn the corner, from the beam into the column. But notice, that even for a fairly coarse mesh, laminar (i.e. regular streamlined) flow is exhibited in the beam. Interestingly, the concavity issue of Fig. 10.6 is corroborated in Fig. 10.7, with tension on the top of the beam near the beam/column interface joint. Also, notice in Fig. 10.7 that the arrows have a magnitude and a direction. The size of the arrows reflects the magnitude of the principal stress in the member, and the direction of the arrow clearly identifies that path of load flow. This powerful tool can be used by designers who want to "let the structure tell them what it wants to do". It has direct links to the major innovation by Professor Jorge Schlaich known as the "strut and tie method", which places reinforcement in line with tensile flow of forces. Finally, note in Fig. 10.7 that the overhanging cantilevers do indeed have stress and load flow, but the arrows, which are drawn to the same scale as the arrows in the central span, clearly show that these are an order of magnitude smaller.

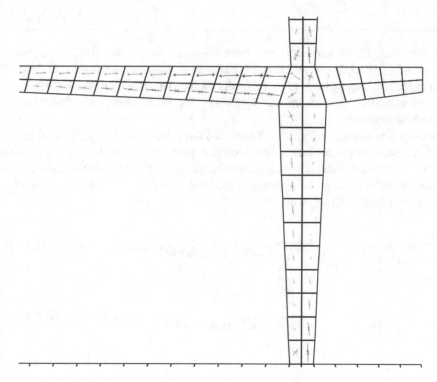

Fig. 10.7 Flow of forces from beam into column

Figure 10.8 shows one half of the frame, the other half exhibiting mirror symmetry. Here, the color coding qualitatively describes the magnitude of the forces, and the direction, which is a bit more difficult to see at this scale, is again noted by the orientation of the arrows.

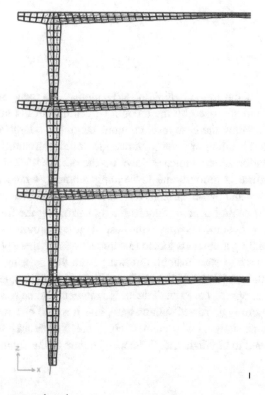

Fig. 10.8 Flow of forces along several stories

The next steps clearly demonstrate the power of linking the tools of the structural engineer, with the aesthetics of the architect. Frame refinement is based on a performance criteria, and the performance here is bending due to gravity loads and bending due to lateral loads. Having the tool of a finite element program such as SAP2000 or Karamba rapidly allows for the refinement of a frame based on these bending moment diagrams.

The first step is to subject the frame to a lateral load that is based on the seismic principles described in Chap. 5. The ease of the method in SAP2000 is remarkable. Figure 10.9 shows the same frame subjected to a lateral load arising from a horizontal acceleration of 0.3 g, where g is the acceleration of gravity. But the entire mass of the structure is accelerated, not just a point load in the plane of the diaphragm as was done in Chap. 8.

Fig. 10.9 Frame subjected to lateral 0.3 g acceleration

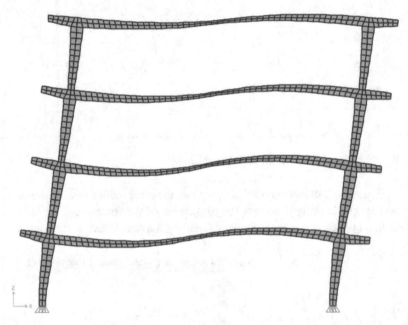

Before any refinement of the frame begins, observe closely the exaggerated deformed shape of Fig. 10.9. Students are often surprised by the amount of bending in the horizontal beams due to a lateral, horizontal load. Certainly the base level columns bend, but a large amount of energy can be dissipated by the controlled bending of beams. Notice the double curvature in the horizontal beams, concave up one half of the beam and concave down on the other half. This has profound structural implications! Since curvature is proportional to bending moment, where the curvature goes from concave up to concave down and passes through zero, this is where no bending moment exists. A very fruitful design insight is to control bending in a frame by narrowing the frame at certain points. For example, if this frame had an extreme tapering at the base of each column, as well as at the central point of the interior span, the bending moment is forced to approach zero at these points. Figure 10.10 shows such a small refinement. An important insight, following from this tapering of the frame, arises from considering the fixity at the very base of the structure. It would be completely incorrect to restrain bending moment in those supports, i.e. to "fix" them. No moment can be reasonably transferred to the ground through the purposefully narrowed column base, thus it would be wrong to actually fix them or to imply fixity through some showy, bulky connection. A much more architecturally and structurally elegant idea is to display and to highlight the "V Shaped" nature of the column, truly demonstrating it as a pinned connection. Figure 10.10 shows the greatly exaggerated deformation of the frame due to Dead Load (self weight) only.

Fig. 10.10 Exaggerated deformation due to dead load of two story concrete frame

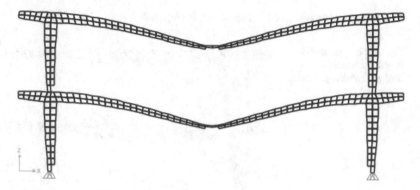

Engineers and architects, students and practitioners, should all be encouraged to draw the bending moment diagram qualitatively directly on the frame. Such a qualitative diagram is shown in Fig. 10.11 Developing this skill takes time and requires checking your hand drawn figures with presumably correct diagrams obtained from a structural analysis program. For concrete construction, it is very convenient to draw bending moments on the tension side of each element. This visually reminds the designer of where reinforcement should be placed. Notice two things in Fig. 10.11: first, the cantilevers hardly bend at all due to self weight, compared to the central span (their individual bending moment diagrams are barely visible). Second, pinning the base of the second story columns to the top of the first story columns removes the transfer of bending from column to column but the continuous, moment carrying connection of the long central span to the column induces a lot of bending, which must be equilibrated by the lower column, not by the pinned upper column.

Fig. 10.11 Qualitative bending moment diagram of stacked three hinged arches subject to dead load

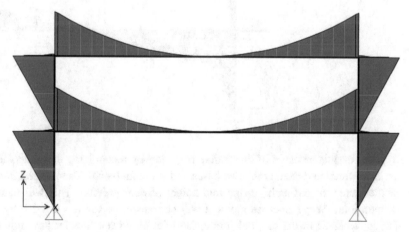

Similar to Fig. 10.9, apply a lateral load of 0.3 g self weight to the frame of Fig. 10.11. The steps to do this in SAP2000 are as follows.

- Define > > Load Pattern > > My03g or some other descriptive name, making the type Other to avoid Dead Load, which seems like a contradiction but it isn't

- Select >> > (select all the areas) and Assign> > Area Loads> > Gravity and choose My03g for the Load Pattern and type in 0.3 for the Gravity Multiplier in the X direction.

The greatly exaggerated deformation is shown in Fig. 10.12.

Fig. 10.12 Exaggerated lateral deformation of stacked three hinged arches as frame

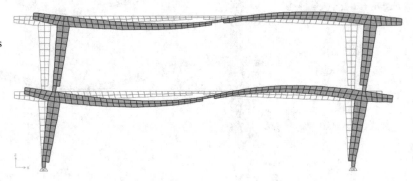

The bending moment diagram for the lateral load case of 0.3 g self weight is drawn in Fig. 10.13 to the exact same scale as the bending moment diagram for Dead Load downwards, previously shown in Fig. 10.11. Notice in the lateral load bending moment diagram of Fig. 10.13, how the first floor interior beam bends significantly more than the second story interior beam. Of course, for Dead Loads, both interior beams bend the same amount.

Fig. 10.13 Qualitative bending moment diagram of stacked three hinged arches subject to lateral load

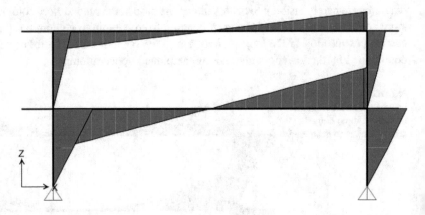

One possible redesign of this frame, is to start by recognizing that it acts like a three hinged arch, i.e. each portal is pinned at the two bases and at the mid-span. Then, redesign by moving the "hinges" of the "three hinged arch" to the mid-height of each column. This will naturally stiffen the beam/column joint. Why? Because moment will be transferred to the second story columns with this new design, whereas in the original frame, the joint could not transfer any moment to the second floor columns because of the inserted hinge there. The deformed shape of the redesigned frame is shown in Fig. 10.14 for Dead Load only.

Fig. 10.14 Refined frame exaggerated deformation due to dead load

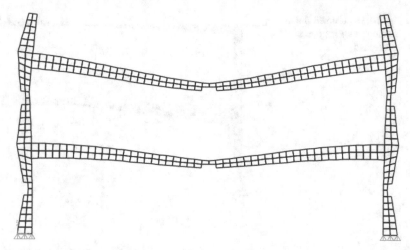

The deformed shape for the 0.3 g is shown in Fig. 10.15 and it is indeed laterally stiffer than the frame of Fig. 10.12.

Fig. 10.15 Refined frame exaggerated deformation due to lateral load

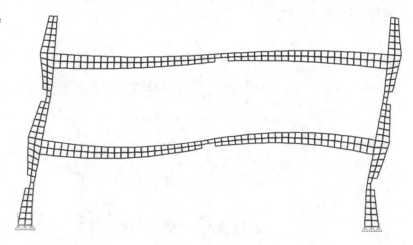

The corresponding bending moment diagram for this 0.3 g lateral load is shown qualitatively in Fig. 10.16.

Fig. 10.16 Refined frame bending moments due to lateral load

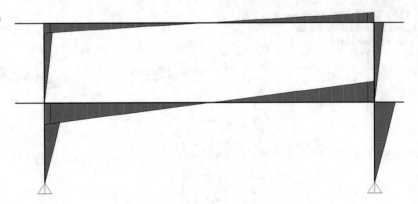

A frame similar to the one of Fig. 10.3 is shown in Fig. 10.17. It has larger cantilevers than before and a slightly smaller interior span. Programmatically, this form might lend itself well to a hotel, as the rooms could be far from each other, each with a good view, and the public central corridor neatly separates the private overhangs. Yet another redesign of this frame would begin with a careful study of the deformed shape, and the resulting bending moment diagrams, for any number of reasonable gravity and lateral load patterns.

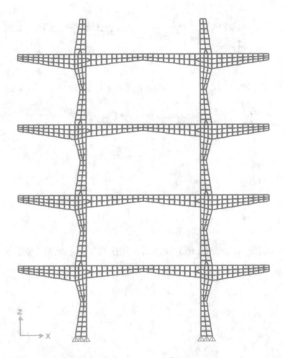

Fig. 10.17 New design with larger overhangs and shorter central span

The deformed shape for Dead Load is shown in Fig. 10.18.

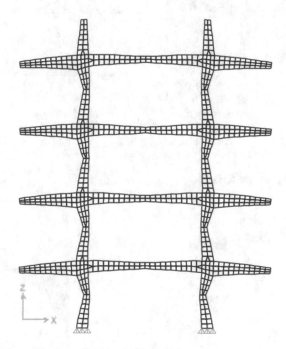

Fig. 10.18 Exaggerated deformation due to dead load of new design

Suppose the frame was to be redesigned to reflect the bending moment diagram, namely, to put material where bending is large and to remove material where bending is small, but this time, the bending will occur from Dead Load Gravity, plus lateral load to the right and lateral load to the left.

Figure 10.19 shows the deformations due to these three loading patterns, as well as the corresponding qualitative bending moment diagrams. Scales of deformed shaped are shown to same degree, which is not strictly necessary, but scales of bending moment diagram are shown to the same degree and that is absolutely needed when redesigning the frame. This allows one to recognize relative magnitudes of bending moments

Fig. 10.19 Three load patterns and bending moment diagrams as possible form generator

Suppose the designer wanted to superpose the bending moment diagrams of Fig. 10.19. The mirror symmetry of the two bending moment diagrams would cancel each other out during such a superposition. This is so, because the simple superposition of the purely positive and negative mirrors of each other would result in a net of zero. A far more subtle and useful technique is that of "enveloping", namely capturing the profile of one loading pattern's bending moment diagram, and capturing the profile of the second, and pasting them both together to create a "bubble" or an enveloping surface that accounts for both. Structural engineers are comfortable with the idea of an envelope, but students rarely are given an opportunity to actually create one. Structural analysis programs are designed to capture envelopes, as well as simple superposition. To capture an envelope in SAP2000, use the following steps:

- DEFINE > > LOAD COMBINATIONS > > ADD NEW COMBO > > then add the Dead Load, and both lateral loads to this combination as shown in Fig. 10.20

Fig. 10.20 Creating an envelope load combination in SAP2000

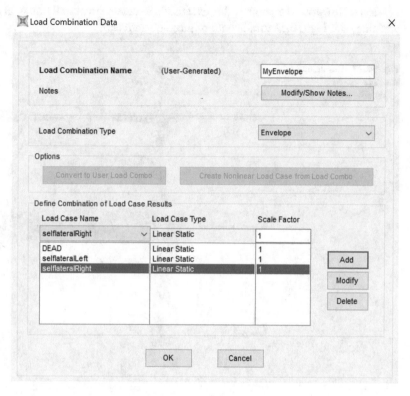

Drawing the envelope bending moment diagram results in a somewhat awkward looking profile as shown in Fig. 10.21

Fig. 10.21 Resultant envelope at awkward scale

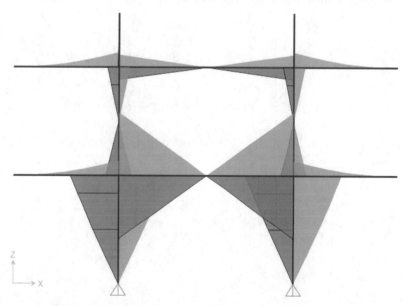

The shape of Fig. 10.21 would not be appropriate for a structural frame as it is far too bulky. But simply scaling down the envelope bending moment diagram results in a more slender profile that might be more suitable. But even the scaled down profile shown in Fig. 10.22 might be too difficult to construct. However, the point of this exercise is to create structurally rational frames that are visually arresting, and Fig. 10.22 fulfills both of those objectives.

Fig. 10.22 Resultant envelope at more reasonable scale

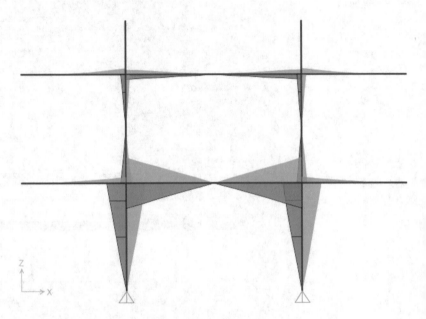

Perhaps the discrepancy between the first floor bending moments and the second floor bending moments is still too large. This leads to the idea of Fig. 10.22, which created a new frame profile shown in Fig. 10.23. Here, the first floor elements are more robust than the upper story elements, but the distinction is more subtle than they were in Fig. 10.22.

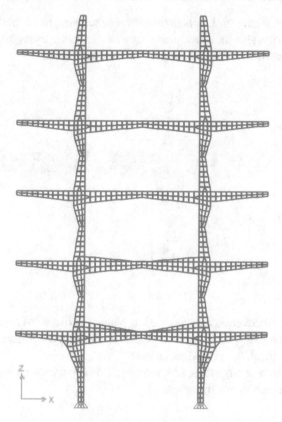

Fig. 10.23 Using envelope to refined previous frame, lower stories bulkier

Another redesign idea could be to recognize that the moment from the outside cantilevers, connecting to the column, is much smaller than the moment from the inside central span to the same column. This was pointed out before. Redesign the frame, such that the large interior beam meets the column with a more substantial connection than does the cantilever into the column. A portion of the frame highlighting the new connection is shown in Fig. 10.24.

Fig. 10.24 Redesigning the column to beam interface

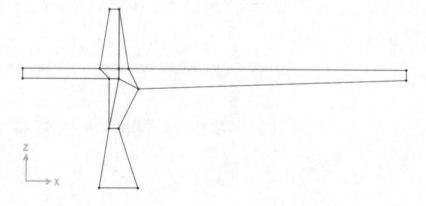

Many plugins such as Weaverbird do beautiful mesh generation, but if meshing "by hand" be sure to remove duplicate nodes and judiciously use skewed elements, trying to keep shapes uniform. Figure 10.25 shows an adequate, but still not perfectly uniform mesh.

Fig. 10.25 Meshing the new interface

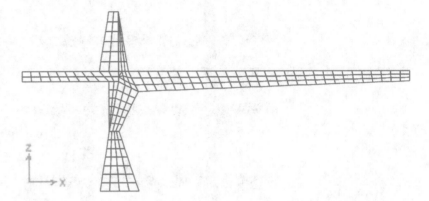

Note, the base in Fig. 10.25 cannot be modeled as pinned, as that would create a mechanism in the column. Furthermore, the base should not be modeled as pinned since it flares as an upside-down V, thus it is "asking to be" fixed, not pinned, at its base.

Mirror and translate to take advantage of symmetry. This should always be done when generating digital forms. Fig. 10.26 shows this operation.

Fig. 10.26 Possible new design with new interface

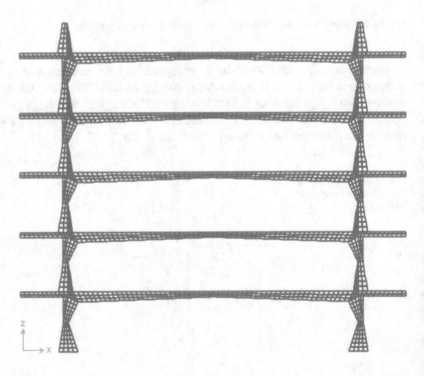

The principal of avoiding a mechanism must be applied to the beams as well, but clearly the beam/column interface was the formal driving idea in this study. In other words, the shape of the frame was dictated by increasing the moment capacity of the column to the interior beam. A secondary formal design idea was increasing the capacity of the smaller exterior cantilever beam to column. The deformed shape, due to Dead Load, of the newly designed frame is shown amplified in Fig. 10.27.

Fig. 10.27 Exaggerated deformation due to dead load of possible new design

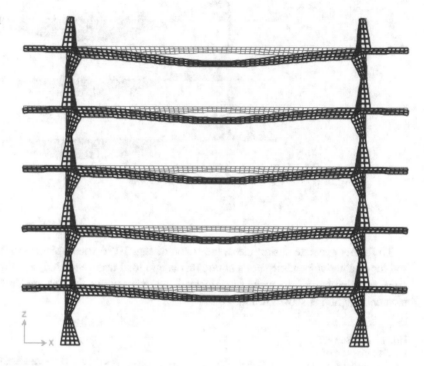

A careful review of the deformed shape in Fig. 10.27 shows something potentially troublesome. That is, the central narrow portion does not act as a hinge. Rather, this central region of the interior beams is decidedly concave up in its deformed shape. This clearly demonstrates that the central region of the interior beam is in fact moment carrying, even though the beam itself is narrowed in cross section at this point. This may cause concern, as the moment is large at mid-span, precisely where the cross section is small. The qualitative bending moment diagram of this frame corroborates the concern, there is quite a lot of bending in the central portion of the interior span. This is shown in Fig. 10.28.

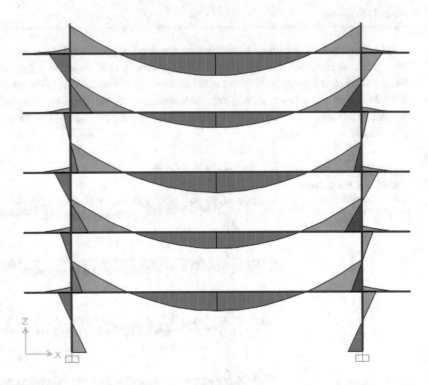

Fig. 10.28 Bending moment diagram shows large moment in central portion due to dead load

To further generate design ideas, the frame of Fig. 10.26 was subjected to 0.3 g lateral static load and the qualitative bending moment due this lateral load was generated, but it was drawn to the exact same scale as before, to make an informed judgment about the design. This lateral load bending moment diagram is shown in Fig. 10.29.

Fig. 10.29 Bending moment diagram of possible new design due to lateral load

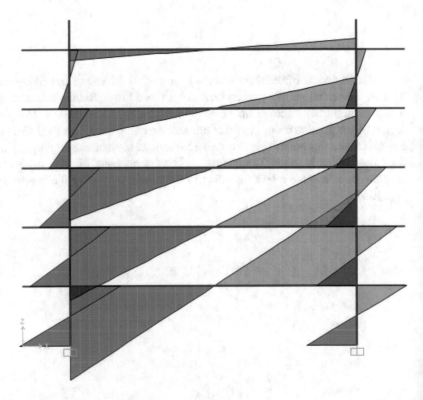

It is decided that although the Lateral Load bending moment is zero at the mid-span point, the Dead Load bending moment there is far too large to be a satisfactory design choice. Thus, the next iteration of design imagines the bending moment distribution if that central node was "hinged", or designed to not carry any bending moment. This is achieved by removing even more material from the central point. Detailing of reinforcement would still require shear to be transferred there. One way of detailing this would be to create an X form with the reinforcing bars over this gap.

The deformed shape for this new design iteration is shown in Fig. 10.30. An interesting, but unexpected feature shown by this deformed shape is the very large displacement of the top story. This arises from the fact that no columns above this topmost beam are present to restrain the structure from bowing inwards. The inwards rotation of the top vertical elements is so large that the second to the top beam actually exhibits concave downward curvature, which is somewhat surprising for Dead Load. That phenomena could be stored away in a sketchbook for future design development, perhaps with some added tensile elements to restrain the inward bowing. For now, focus on further studies of the present frame.

Fig. 10.30 Exaggerated deformation due to dead load of subsequent design with central hinge

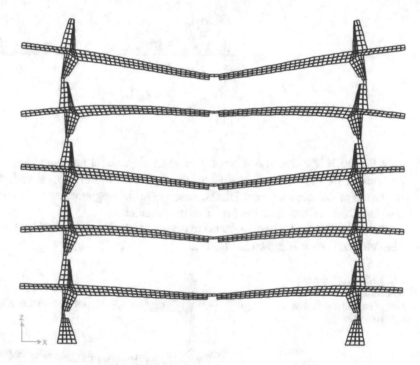

Draw the qualitative bending moment diagram of the frame in Fig. 10.30, but do so to the exact same scale as before, to make an informed judgment about the design. This diagram is shown in Fig. 10.31.

Fig. 10.31 Bending moment diagram due to dead load of subsequent design with central hinge

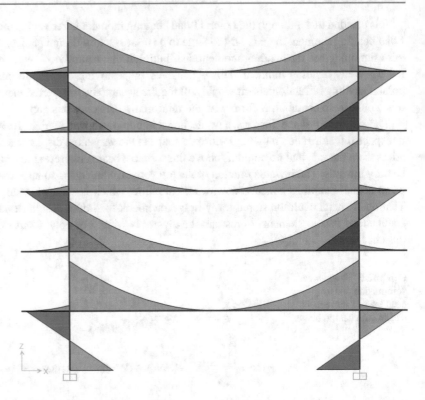

In Fig. 10.31, the Dead Load bending moment diagram for the central beam is still a parabola with the familiar $wL^2/8$ magnitude, yet the overall diagram position changes with the changing boundary conditions of the interior beam. That is, there are restraining moments on either side of this beam, as was described in Chap. 6 and in Fig. 6.26 in particular.

The design of Fig. 10.30 is subjected to 0.3 g lateral self weight static load, as was done previously. The deformed shape is shown in Fig. 10.32.

Fig. 10.32 Exaggerated deformation due to lateral load of subsequent design with central hinge

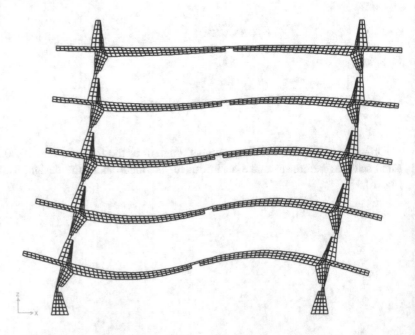

Notice the dramatically larger curvature in the lowest beams of Fig. 10.32.

The bending moment diagram due to lateral load is unchanged from the previous case of no central hinge on the interior beam. As predicted, the lowest beams have the most moment by far.

Next, dig deeper into the stiffness ratios of beam to column. Consider only one floor and one bay of the previous frame, and use this as a baseline. Figure 10.33 shows the baseline structure to be studied.

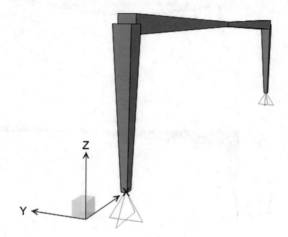

Fig. 10.33 Study one bay only with new design idea

Draw the deformation for this structure due to dead load. Notice the switch from concave up curvature, to concave down curvature, which happens at approximately ¼ of the span away from the vertical columns. This is shown in Fig. 10.34

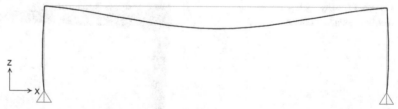

Fig. 10.34 Exaggerated deformation of new design idea due to dead load

The bending moment of this single portal frame, due to Dead Load only, is shown in Fig. 10.35. As predicted, the diagram for the horizontal beam is indeed parabolic, with the familiar $wL^2/8$ peak, but it is restrained by the connection to the column at either end, which shifts the entire parabola.

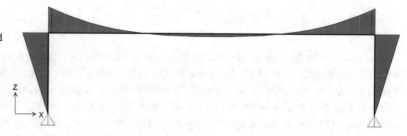

Fig. 10.35 Bending moment diagram of new design idea due to dead load

As before, subject the frame to some convenient static lateral load, here 0.3 g was used once again. The deflected shape is as expected, with concave up switching to concave down curvature at the midpoint. This is shown in Fig. 10.36, along with the corresponding qualitative bending moment diagram.

Fig. 10.36 Deformation and bending moment diagram of new design idea due to lateral load

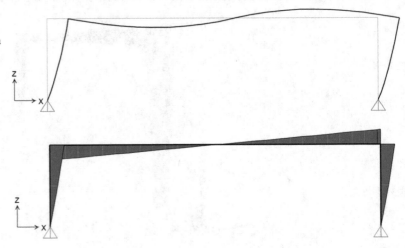

Now make the design decision to modify the frame by leaving the beam as is, but refining the column to be very slender. The refined portal frame is shown in Fig. 10.37. The motivation behind this design change is somewhat aesthetically driven, but the consequence will be an increased bending moment in the central portion of the beam, and a reduced bending moment in the columns. Engineering skills will help to intuit why this is so.

Fig. 10.37 New design idea with slender column

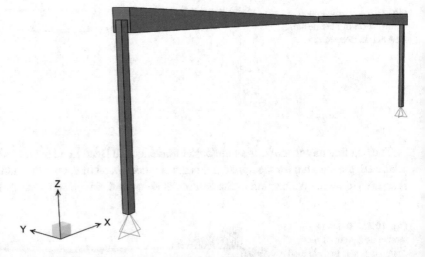

The reason why the energy of bending shifts away from the columns and moves towards the central span is made clear in Fig. 10.38. The main horizontal beam still feels a parabolic bending moment with a peak of $wL^2/8$, but notice how the entire parabolic shape is shifted compared to the bending of the frame in Fig. 10.33. The beam to column interface for the frame of 10.37, with the slender column, simply cannot transfer the moment adequately, because the column is too flexible, not because the beam isn't stiff enough. Comparing the two diagrams side by side drives home this point.

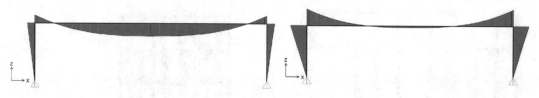

Bending of frame in Figure Frame 10-37   versus   Bending of frame in Figure 10-33

Fig. 10.38 Comparing bending of two different column configurations in single bay

For lateral load, there is no perceptible difference in bending moments for this second case of a slender column of Fig. 10.37, compared to the previous configuration of Fig. 10.33. Figure 10.38 clearly demonstrates the engineer's version of the famous architectural dictum made famous by Louis Sullivan, but somewhat redacted by the general public. Sullivan stated "Form ever follows function", but Professor David Billington flipped the statement around by saying "function follows form" rather than the popular "form follows function". By this, Billington meant that function, such as the magnitude of bending, or how a frame functions, deforms and stresses, is directly linked to the form of the frame. Figure 10.38 clearly demonstrates "function follows form". Professor Zalewski and Professor Allen made the phrase their own by linking "form and forces". Their point was that the forces within a structure are directly channeled and guided by the form of the structure.

Students were prompted to use the technique of the bending moment diagram envelope to refine frames supporting a three story building. Figures 10.39 and 10.40 show samples of the student work. Part of the project prompt was to assess their graphic design skills via a concise poster that demonstrates their design thinking. Such graphic communication is an important skill for building designers, who need to rapidly convey their ideas to potential clients.

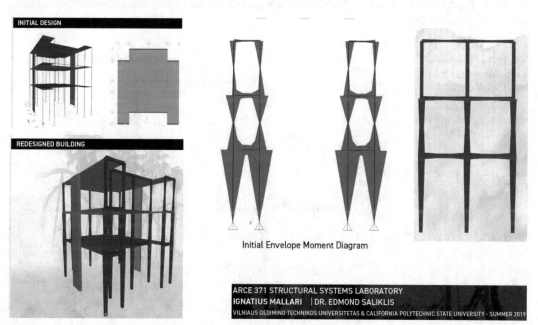

Fig. 10.39 One student project using bending diagrams as form generator

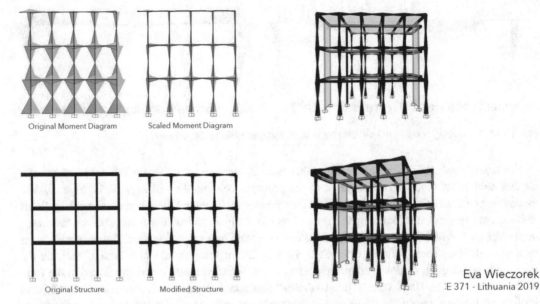

Original Moment Diagram   Scaled Moment Diagram

Original Structure   Modified Structure   Eva Wieczorek
                                                                                  :E 371 - Lithuania 2019

Fig. 10.40 Another student project using bending diagrams as form generator

A fairly new and very powerful tool available to all is the plugin Karamba, which works with Grasshopper in the Rhino environment. The principles described in this chapter, namely investigating the shape of the bending moment diagrams for gravity and for lateral loads, and then using those shapes as the form generator for a refined frame, were all programmed in Grasshopper and analyzed in Karamba. Figure 10.41 shows a typical Karamba generated bending moment diagram for a frame subjected to gravity load.

Fig. 10.41 Bending of   Programmed by Joshua Lange
frame due to dead load
studied in Karamba

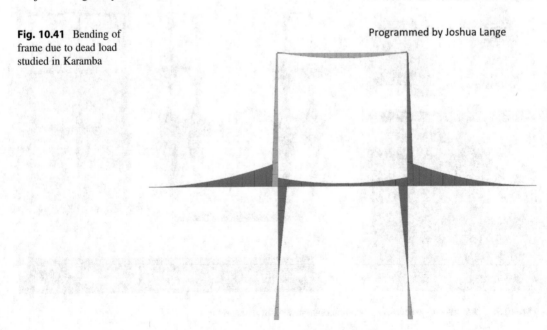

The same principle of accelerating the self-weight at some fraction of gravity, for example 0.3 g, was used in Karamba. The resulting bending moment diagrams for lateral loads are shown in Fig. 10.42.

Fig. 10.42 Bending of frame due to lateral load studied in Karamba

Grasshopper/Karamba then was used to overlay or "envelope" the individual bending moment diagrams. This stacking within Grasshopper and Karamba is shown in Fig. 10.43 and then as output in Rhino in Fig. 10.44.

Fig. 10.43 Envelope of
three load patterns
using Karamba

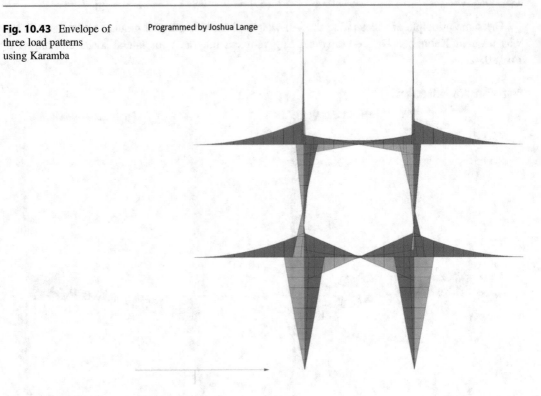

Programmed by Joshua Lange

Fig. 10.44 Superposition
of bending moment
diagrams to create frame
form in Rhino from
Karamba

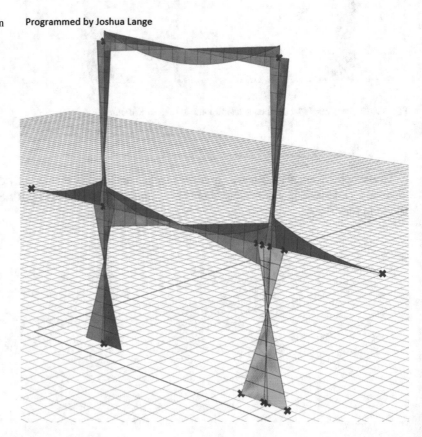

Programmed by Joshua Lange

For those totally unfamiliar with Grasshopper and Karamba, Fig. 10.45 shows a piece of the script used to generate this frame refinement. It is an object oriented language that uses powerful and succinct icons, rather than lengthy text. Icons, which are primarily Python based small subroutines, have distinct and strictly defined graphical inputs and output, much like a wiring diagram.

Programmed by Joshua Lange

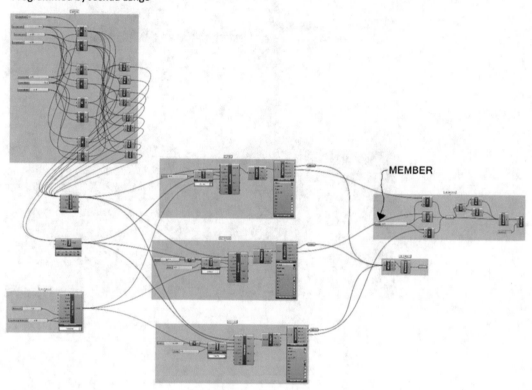

Fig. 10.45 Sample Grasshopper script to generate form

Grasshopper and Karamba both work within Rhino, which has tremendous rendering tools. Fig. 10.46 shows the theoretical refined frame, created in Karamba but rendered in Rhino.

Fig. 10.46 Rendered form based on bending moment diagram

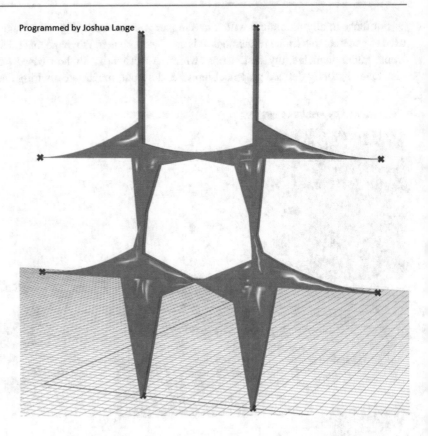

Shells in the Studio

Brilliant structural engineers such as Pier Luigi Nervi and Eduoardo Torroja have convincingly argued that an understanding of the flow of forces in structures from a mathematical point of view can lead to aesthetically elegant structures from a design perspective. In an interdisciplinary studio that combined architecture students and faculty with architectural engineering students and faculty, there was a tremendous opportunity for not only linking such left brain and right brain thinking, but to also to triangulate the creative process by adding craft of construction to the design problem.

In a fourth year Interdisciplinary Design Studio, architectural engineering students were paired with architecture students, with the goal of blending art and technology. The students were asked to negotiate the challen
ges of form finding based on solely programmatic concerns with form finding based on mathematical, compression-only algorithms. Extremely thin, reinforced concrete shells were the centerpiece of each studio. Professor Ansgar Killing and Professor Clare Olsen, both of the Architecture Department at Cal Poly, were instrumental in designing the studio, and mentoring the students. Without them, this studio would never have come to fruition. We worked together to establish the following parameters, which were guiding principles of the Interdisciplinary Studio:

1. A precedent study of significant thin shell reinforced concrete structures from a load flow point of view. This same technique was described in Chap. 9.
2. To Design, Analyze and Build a compression-only large scale model of an ultra-thin reinforced concrete shell.
3. Students had constrained freedom in choosing the site for their virtual shell, constrained freedom in determining the programmatic spaces of their shell, constrained freedom in the erection schemes for the on-campus model as well as for the site-specific virtual shell, constrained freedom in the structural analysis.

There were two philosophical reasons for focusing on the problem of thin shell concrete structures in this Interdisciplinary Studio. The first was that thin shell structures are somewhat unique, in that for even an elementary understanding of these architectural wonders, much less mastery of them, they require deep immersion in three phases of planning: the architectural design, the structural analysis, and the constructability of the shell and of the shell's landing points to the ground. This tripartite approach was an ideal fit for an undergraduate interdisciplinary studio because it drew from a varied set of skills—no single discipline could execute such a project in a studio, several disciplines were needed.

© Springer Nature Switzerland AG 2020
E. Saliklis, *Structures: A Studio Approach*, https://doi.org/10.1007/978-3-030-33153-5_11

The second reason was slightly less tangible. In various great works of architectural engineering, the line between the Engineer and Architect was blurred. Felix Candela was a prime example of this porous boundary between Designer and Analyst. Candela also embodied the third leg of the tripartite triangle, namely he was a Builder as well. This immersion into a thin-shell studio was a small attempt to rekindle that spirit of Designer/Analyst/Builder. Immersing the students in an experience that drew from their own specialty, but required them to negotiate and communicate with non-specialists, fostered the development of the Designer/Analyst/Builder.

Given that the studio focused on shells, the choice of "compression-only" made the most sense, because the tools to find a 3D funicular are now readily available. To facilitate physics-based funiculars, tutorials were provided in Grasshopper/Kangaroo/WeaverBird and students used this process to create shells that appear to be freeform, but in fact, are strictly mathematically constrained to have no bending. This theoretically would allow for impossibly-thin structures to be created, were it not for buckling concerns.

The following figures are of very detailed precedent studies of a shell, via SAP2000 as well as via Grasshopper and Karamba. Figure 11.1 shows a load flow analysis of Felix Candela's Chapel Lomas de Cuernavaca. It is well known that Candela was the master of the hyperbolic paraboloid, and Cuernavaca is wonderful example of an extremely dramatic hyperbolic paraboloid. The question that was asked by the analysis in Fig. 11.1 was whether or not the path of compression was along the meridians? Also, what of the concave up portion in the center of the shell, was it in tension? The load flow arrows, generated by SAP200, are shown in Fig. 11.1. These load flow arrows are similar to the load flow images in Figs. 10.7 and 10.8. The work shown in Figs. 11.1 through 11.9 were all programmed by Geoffrey Sanhueza.

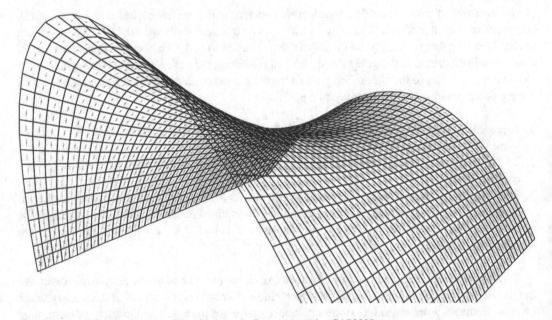

Fig. 11.1 Load flow in Candela's Chapel Lomas de Cuernavaca using SAP2000

Inspection of Fig. 11.1 shows compressive flow, that is uniform, analogous to laminar flow in fluid mechanics, down along the main arch-like forms that define the shell in the meridians. Very small tension is found in the concave upward portion of the saddle shape. This is to be expected because the hyperbolic paraboloid is not a funicular, it would not be compression-only structure. Candela chose this form partly because of its aesthetic power, but mostly because of its constructability. Such a form can be generated by straight lines, thus no curvilinear formwork was used by Candela, only straight, dimensional lumber.

Figure 11.2 shows the laminar, compressive, downward flow of forces along the arches, to the ground.

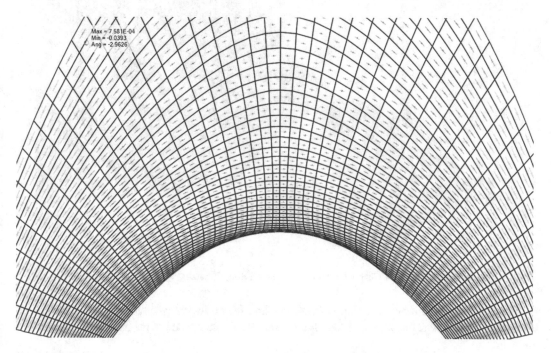

Fig. 11.2 Front view of Load flow in Candela's Chapel Lomas de Cuernavaca using SAP2000

Figure 11.3 shows the same Cuernavaca Chapel, but this time the analysis is performed within Karamba. Contour lines describe principal stresses, with red lines depicting the worst compression and blue lines depicting the least-worst compression. In structures, these two principal stresses are always perpendicular to each other. The direction of the contour lines matches the direction of load flow, as did the arrows in SAP2000. But the contour lines exhibit magnitude by the density of the line spacing, closely packed lines means a more extreme value. Karamba shows the blue lines, namely the least worst compression, perpendicular to the tightly spaced red compression lines. This corroborates SAP's analysis, but these images are more visually arresting.

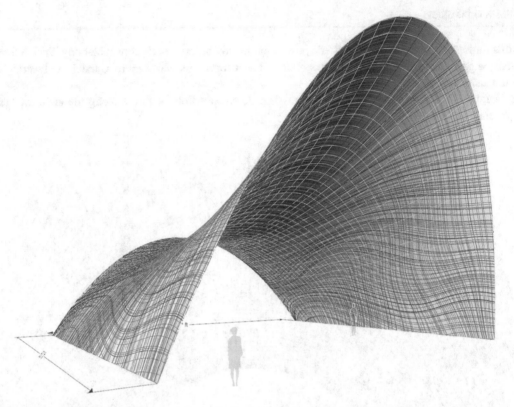

Fig. 11.3 Load flow in Candela's Chapel Lomas de Cuernavaca using Karamba

Figure 11.4 is a similar view to Fig. 11.2, but this time the results are generated by Karamba. Notice the loosely spaced blue line contours, signifying a smaller absolute value of magnitude of stress in that direction.

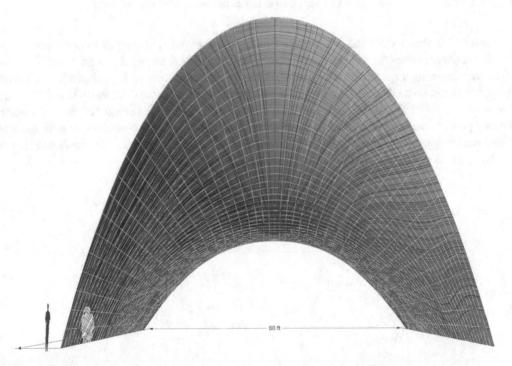

Fig. 11.4 Front view of Load flow in Candela's Chapel Lomas de Cuernavaca using Karamba

There were two innovative features of this studio. The first is to require from undergraduates, the computational approach to finding a funicular, three dimensional structure. A funicular in two dimensions could be thought of as a hanging chain, it is in pure tension when hanging freely, regardless of the support conditions or the level of static indeterminacy. These funicular were analogous to nets made up of chains, but the nets are subjected to upwards gravity. The second unique feature of the studio is the construction of model shells using real reinforced concrete. Constructing such physical shells requires time, effort and financial support.

An intriguing three dimensional, funicular form is shown in Fig. 11.5. It was generated by Grasshopper in conjunction with Kangaroo 2, and then the stress analysis was performed in Karamba. The stress analysis allows for load flow to be shown as contour lines, density of lines corresponding to magnitude of load, orientation of lines corresponding to direction of load. Red denotes the worst principal compressive stress, which is best thought of as 9:00 o'clock on Mohr's Circle, and blue is best thought of as 3:00 o'clock on Mohr's Circle, i.e. the least amount of compression. Recall that this is a funicular, so it will be compression only under self weight.

Fig. 11.5 Typical funicular shell generated in studio

Figure 11.6 shows the first steps of a Grasshopper script to create a 3D funicular. This is only one way of deriving a funicular, 3D shape. Of course, each designer could come up with alternate ways of finding such a shape, some might be more efficient, some might have more flexibility in their user friendliness. This is simply one way of finding the form.

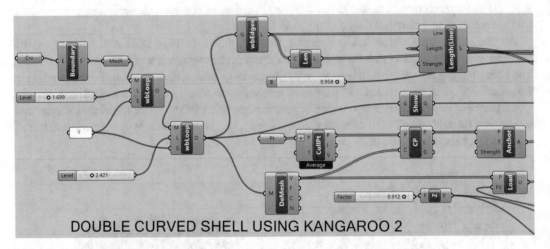

Fig. 11.6 Beginning of Grasshopper script to create funicular shell

The Grasshopper script reads from left to right, with the inputs of icons being on the left of each and the outputs being on the right. Initially, a boundary line for the footprint of the shell is established, drawn in the Rhino environment as some polygon. The CRV icon allows Grasshopper to recognize the Rhino entity. Geometry of the edges is manipulated such that points from lines are culled, this allows for the establishment of anchor points, which are the boundary supports connecting the shell to the ground. Figure 11.7 shows the next steps.

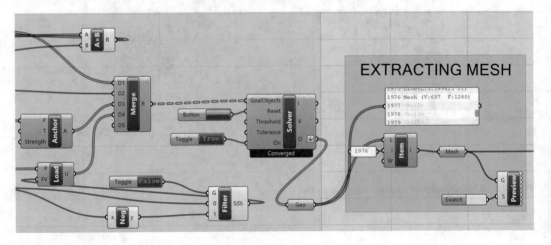

Fig. 11.7 Middle of Grasshopper script to create funicular shell

Merge is a convenient icon that simply takes multiple inputs and reorganizes them into a single stream that can be read in an orderly manner by Kangaroo. Noteworthy is the SOLVER icon in Fig. 11.7. This is not a stress analysis on the shell, rather, it generates the funicular from the net, or grillage of chains. It can be lofted to any height the designer chooses.

Here in Fig. 11.8, a geometric mesh is switched into a finite element mesh of area elements. Although one can think of the form generation as coming from a hanging net subjected to reverse, or upwards gravity, the net is used to ultimately form monolithic, continuous elements which are bounded by the lines of the net. This is done by MeshToShell in Karamba.

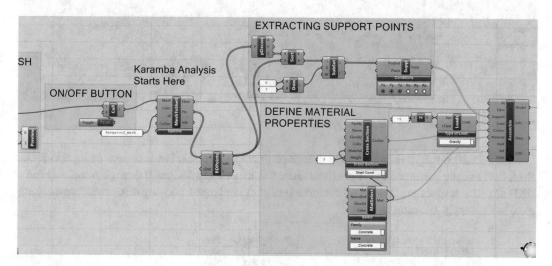

Fig. 11.8 End of Grasshopper script to create funicular shell

Figure 11.8 includes steps that are in the comfort zone of engineers and engineering students, but that are decidedly new to architects and to architecture students. Here, boundary conditions are created from the so-called geometric anchor points. Material properties and cross-sectional properties are provided for the finite elements. The force vector is created and the stiffness matrix is assembled. Note in Fig. 11.8 that the forces now are applied to the shell downwards, in the −Z direction, these are the gravity loads on the shell, not the lofting force needed to form the funicular.

Figure 11.9 shows some of the post-processing features available in Karamba, which become images in Rhino. The most impressive images arise from the contours describing principal stress. Stress is directly indicative of force, and as described briefly before, the principal stresses are 90° apart in the physical world and 180° apart in Mohr's Circle which is a graphical tool that describes principal, or worst, stresses. Buckling is Eigen buckling. Eigen buckling is of limited value in the study of shells, since real buckling loads will never reach theoretical Eigen buckling upper bounds, because of the inevitable presence of material and geometric imperfections in real shells. Nevertheless, it is a useful tool for estimating critical loads. Buckling is of great concern as the shells become ultra-thin.

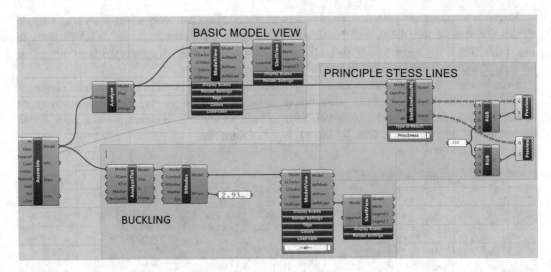

Fig. 11.9 Postprocessing script which uses Karamba

Figure 11.10 displays the previously described principal stresses, but this time using SAP2000. Here, the funicular generated by the previous script was exported to SAP2000 through AutoCAD and a DXF file. The stresses are indeed all compressive, with the largest non-negative stress approaching zero, but narrowly avoiding tension.

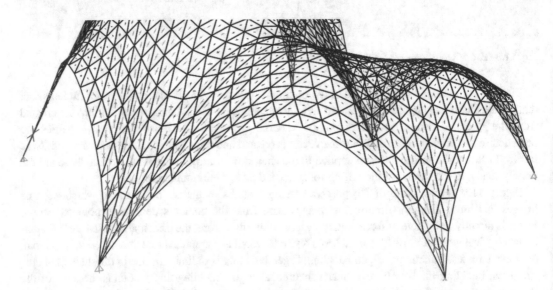

Fig. 11.10 Kangaroo2 shell exported to SAP2000 for load flow study

Figure 11.11 summarizes the ideation process from the broad brushstrokes of programmatic concerns, through a digital, compression only monolithic shell.

Design

The design of shells in the studio responded to both architectural, programmatic needs as well as to structural concerns and to construction viability. The following figures show various student responses to the studio prompt. The prompt required initial sketches that accommodate the architectural program, and corresponding, digitally produced, compression-only shells that cover and define the architectural spaces. Figure 11.12 shows a design for a proposed Onsen, or Japanese Bath House.

Design by Erika Meller, Munenari Hirata, Zachary Price and David Nunez

Tapas bar
and Seatings   Skate bowl

Two Main Program   Combine and Expand   Changing Height

Fig. 11.11 Overview of form finding in studio

Design by Geoffrey Sanhueza, Alex Fernandez, Divyash Agrawal and Zak Collin

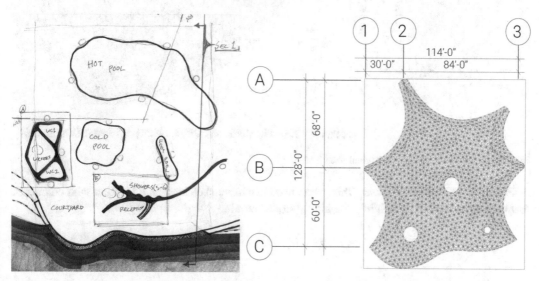

Fig. 11.12 Onsen ideation to shell form finding

Figure 11.13 shows the design for a winery tasting room and educational center. The complexity of the architectural program called for several discrete shells.

Design by Cameron Adelseck, Jeff Kim and Kiana Underwood

Fig. 11.13 Winery ideation to shell form finding

Figure 11.14 is an event center which could be used as a family fun pool space during the day, but could host receptions in the evening in a larger, connected area.

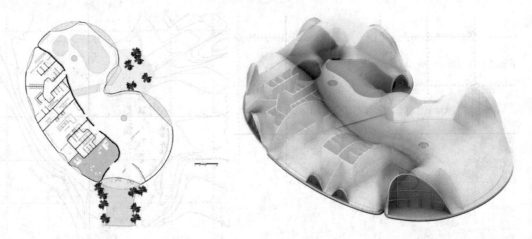

Design by Cynthia Renteria, Ana Lopze, Hennry Garcia and Letty Lopez

Fig. 11.14 Event center ideation to shell form finding

Figure 11.15 is another Onsen, this group used extensive sketching, and ideation with cutouts of foam sheets to nestle the shell into the highly sloped terrain.

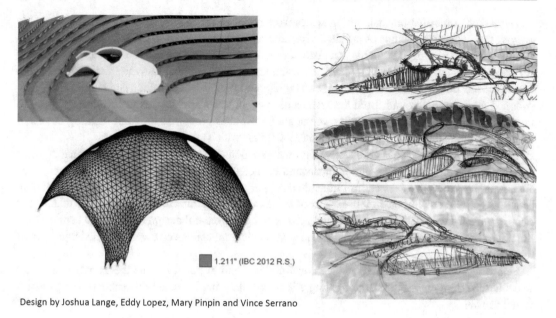

Design by Joshua Lange, Eddy Lopez, Mary Pinpin and Vince Serrano

Fig. 11.15 Spa retreat ideation to shell form finding

Figure 11.16 is another Onsen, with clearly separated spaces for men and for women.

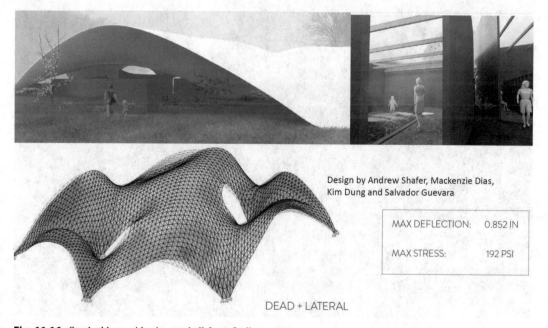

Design by Andrew Shafer, Mackenzie Dias,
Kim Dung and Salvador Guevara

MAX DEFLECTION:	0.852 IN
MAX STRESS:	192 PSI

DEAD + LATERAL

Fig. 11.16 Spa bathhouse ideation to shell form finding

Two other, somewhat unusual design issues were raised in the Interdisciplinary Shell Studio. One was lighting around the oculus of the shell. The students were prompted to try and re-create a "James Turrell" effect of placing LED lights just inside the lip of the oculus. One result of this effect is that the sky appears dramatically highlighted, when viewed through the oculus, from the interior of the shell. Some students pursued this effect further, and highlighted the main body of the shell with LEDs, as it swooped down to the ground. Both the architectural engineering students, as well as the architecture students were fascinated by this portion of the studio. Neither group had explored lighting in such a hands-on manner, and very few were previously familiar with Turrell.

The second issue was the careful study of how the shell landed on the ground. The instructors joked that "transitions are difficult; in life, in music and in shells". There is truth in the joke however. Very often in architecture, one sees clumsy transitions, either around corners, or at the point of negotiation between roof and column or, in this case, roof to ground. The architectural engineering students had some experience with detailing of structural members, i.e. the careful design of structural connections. But none of the students had considered the deep aesthetic implications of a well designed transition of shell to ground.

Figure 11.17 shows a student designed construction detail of how to create the Turrell effect. The students then digitally modeled the lighting effect, and they implemented this scheme in a physical model as well.

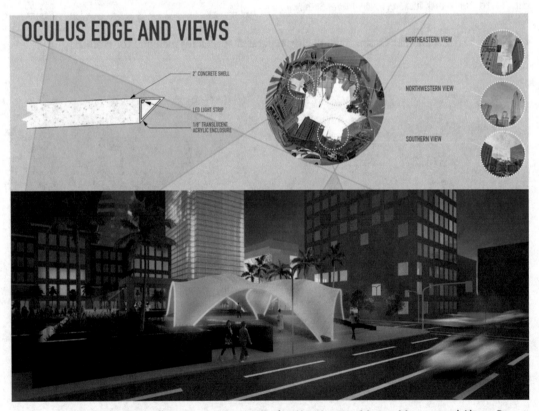

Design by Paris Allen, Dewan Hoque, Keslyn Huntington, Megan Morgan and Alyssa Parr

Fig. 11.17 Lighting detail and digital nighttime lighting study

The carefully designed recess, that allows the LEDs to be hidden from view are shown in Fig. 11.18. Figure 11.18 also hints at the Turrell effect that was desired.

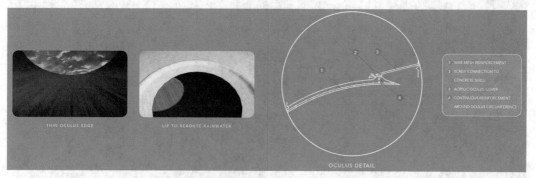

Design by James Blanchard, Lillian Cao, Nick Horaney, Hunter Mosier and Cory Peterman

Fig. 11.18 Different lighting detail and evening lighting study

Figure 11.19 shows a construction detail that is recessed from below, the exact opposite of Fig. 11.18. Here the desired optical effect of the sky was digitally modeled.

Design by Andrew Shafer, Mackenzie Dias,Kim Dung and Salvador Guevara

Fig. 11.19 Another detail and twilight lighting study

Figures 11.20 and 11.21 show nighttime views of the lighting in the model shells. Even at a small scale, the lighting creates a powerful spectacle through the oculi.

Fig. 11.20   More physical models lit up in the evening

Fig. 11.21   More physical models light in evening

Some lighting schemes combined passive daylight conditions, with controlled LED nighttime lighting. Figure 11.22 shows digital mockups of such a design.

Passive Daylight   Colored LED Night Light

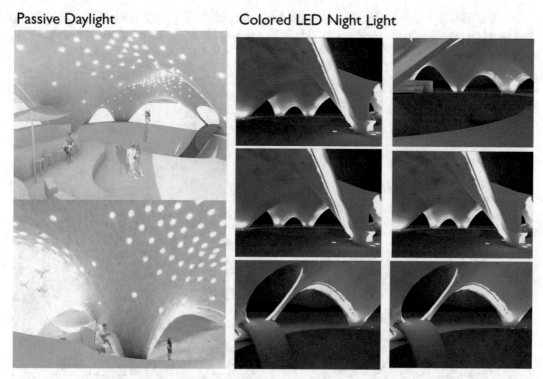

Design by Erika Meller, Munenari Hirata, Zachary Price and David Nunez

Fig. 11.22 Passive daylight study coupled to evening lighting study

Transitions between the shell roof and the ground were given attention. Students physically mocked up details in 3D, but mostly relied on digitally modeling this transition. Figure 11.23 shows a detail of how the shallow space between the sloped roof and the ground might be used.

Design by Geoffrey Sanhueza, Alex Fernandez, Divyash Agrawal and Zak Collin

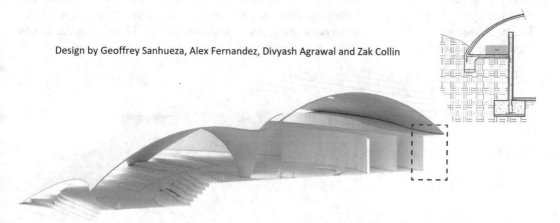

Fig. 11.23 Study of how shell meets ground on steep slope

Another transition between the shell roof and the ground was highlighted with a Corten steel fitting that snugly wrapped onto the shell on one side, with a prominent flange on the other side of the shell. This is shown in Fig. 11.24.

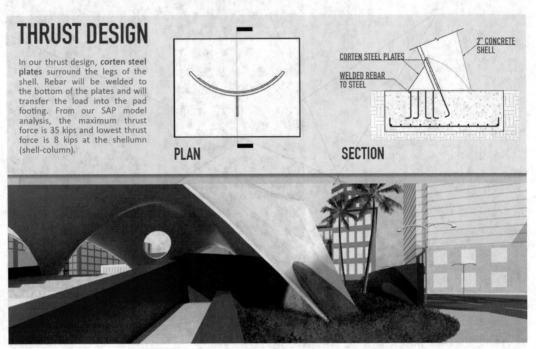

Design by Paris Allen, Dewan Hoque, Keslyn Huntington, Megan Morgan and Alyssa Parr

Fig. 11.24 Suggested Corten steel detail for transition of shell to ground

A particularly fun transition was designed by students for a proposed skate park. They decided to lower one oculus so that skaters could skate through the opening, and descend into the bowl. The overall design is shown in Fig. 11.25 and proposed concrete detailing of another transition where the shell could become a bench before meeting the ground, is shown in Fig. 11.26. This detail was quantitative and used the concrete Code for guidance.

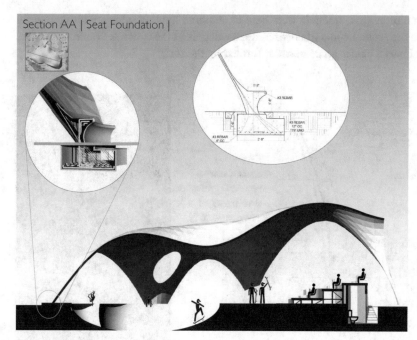

Fig. 11.25 Overview of shell transition to ground

Project by: Erika Meller,
Munenari Hirrata, Zach Price
and David Nunez

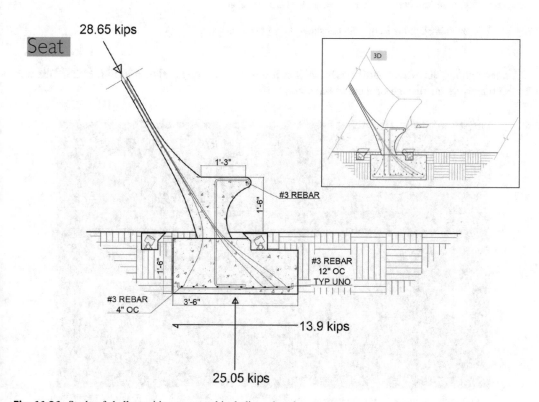

Fig. 11.26 Study of shell transition to ground including a bench

With some of the studio designs, students were able to combine lighting effects with thoughtful transitions of the shell to the ground. Quantifiably reasonable details of the concrete connections were required. Figure 11.27 shows a below grade corridor formed by the shell.

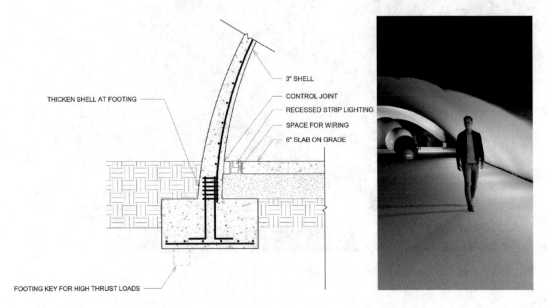

Design by Jackie Anaya, Alex Buchanan, Wood Cheng and Nathan Lundberg

Fig. 11.27   Combining lighting study with transition study of shell to ground

Under-lighting thrown upwards on the shell as it meets the ground is shown in Fig. 11.28. This was for the underground storage of wine at the virtual site.

Design by Cameron Adelseck, Jeff Kim and Kiana Underwood

Fig. 11.28   Lighting study of shell creating winery storage space

Another shell student project was a museum space which had multiple shell segments converging to a focal point. The interstices between the shell segments became dramatic sources of light. The students clearly paid homage to Candela in Fig. 11.29, with the small perforations of the shell, and with the borrowing of ideas from Candela's Bacardi Rum Factory.

Design by Sergio Ochoa, Bryan Boozari, Hailey Rose and Diana Villanueva

Fig. 11.29 Controlled daylight study of shell perforations

Figure 11.30 shows a student project with daytime lighting infusing the transition of shell to ground. This became an inviting seating area, convincingly demonstrated with what the instructors joked was a "B+ rendering".

Design by Andrew Shafer, Mackenzie Dias, Kim Dung and Salvador Guevara

Fig. 11.30 Study of daytime lighting infusing the transition of shell to ground

Finally, construction of models was undertaken in the Interdisciplinary Studio. Three methods of construction were used to create the formwork:

- A waffle grid of laser cut cardboard
- Earthen molded mound
- Inflated plastic

Experiments were conducted with a fourth construction method, namely that of pre-casting individual parts of the shell which were then grouted together.

The waffle grid of laser cut cardboard is well within the comfort zone of architecture students. It was not an extremely attractive option for the instructors however, as it is a bit wasteful with materials and it can only be translated to full scale via cut plywood, which is even more expensive. The earthen molded mound has been used on occasion in the 1960s to build real shells full scale, and certainly molded earth has been used as formwork since antiquity. We like to think of earthen formwork as an ultra low-tech construction method of an ultra high-tech design. Interestingly, the history of inflatable or pneumatic formwork also has a long history. As early as the 1930s, experiments were done in Italy on pneumatic formwork for heavy civil works such as water lines. Walter Neff patented an inflatable formwork system in 1948. D. Bini built hundreds of shells with pneumatic formwork in Italy, and H. Heifitz built many shells with pneumatic formwork in Israel.

In the Interdisciplinary Studio, inflatable formwork was created from the 3D funicular shapes, then the shapes were "flatpacked" or "unrolled" in Rhino to form 2D cutting patterns. Simple thin plastic sheets were then cut out, tracing these 2D forms. They were heat seamed together with a small household iron. The air pressure needed to inflate these forms is very low, on the order of 25 psi (0.17 MPa). But a large volume of air is needed, even for the models created in this studio.

Regardless of what type of construction method was used, the students presented their proposed schemes via drawings and public critiques. The students were also prompted to plan on methodical removal of formwork, commonly referred to as "de-centering".

Figure 11.31 shows a typical "waffle grid" construction scheme. The students anticipated sliding out the dimensional 2×3 lumber to aid in rapid removal of the formwork. Their resultant shell model is shown in Fig. 11.32.

Fig. 11.31 Exploded axonometric drawing of one construction design scheme

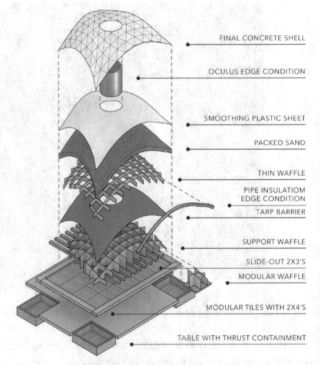

FINAL CONCRETE SHELL

OCULUS EDGE CONDITION

SMOOTHING PLASTIC SHEET

PACKED SAND

THIN WAFFLE

PIPE INSULATIOM EDGE CONDITION

TARP BARRIER

SUPPORT WAFFLE

SLIDE-OUT 2X3'S

MODULAR WAFFLE

MODULAR TILES WITH 2X4'S

TABLE WITH THRUST CONTAINMENT

Design by Jackie Anaya, Alex Buchanan, Wood Cheng and Nathan Lundberg

Fig. 11.32 Resulting built shell from Fig. 11.31 construction idea

Figure 11.33 shows a another "waffle grid" construction scheme. Notice how the waffle grid does not extend down to ground, but rather, is supported by lightweight struts.

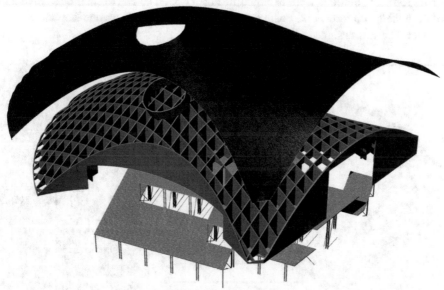

Design by Cameron Adelseck, Jeff Kim and Kiana Underwood

Fig. 11.33 Exploded axonometric drawing of another construction design scheme

The same group realized they needed a fairly hefty tension ring at the base of the shell, to contain the thrust. Their model is shown in Fig. 11.34 along with the tension ring.

Fig. 11.34 Resulting built shell from Fig. 11.33 construction idea

Pneumatic formwork solutions were explored in Fig. 11.35. The actual shell only covered the top portion of the bubbled up formwork. Foam pool noodles were cleverly used to stop the concrete from flowing further than was designed.

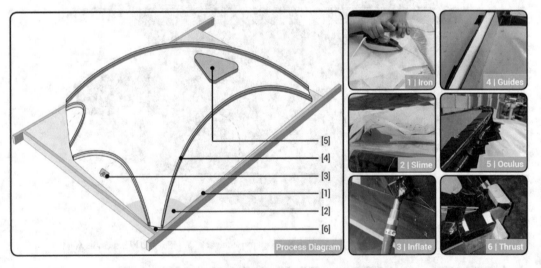

Design by Mark Luzi, Karla Santos, Josh Cannon, Erika Davis and Nora Villanueva

Fig. 11.35 Construction sequence for inflatable formwork

Figures 11.36, 11.37, 11.38 and 11.39 show the multiple steps used to create the extremely thin reinforced concrete shell using this delicate pneumatic formwork.

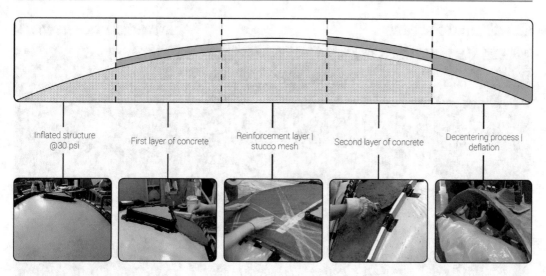

Inflated structure
@30 psi

First layer of concrete

Reinforcement layer |
stucco mesh

Second layer of concrete

Decentering process |
deflation

Fig. 11.36 Details for inflatable formwork

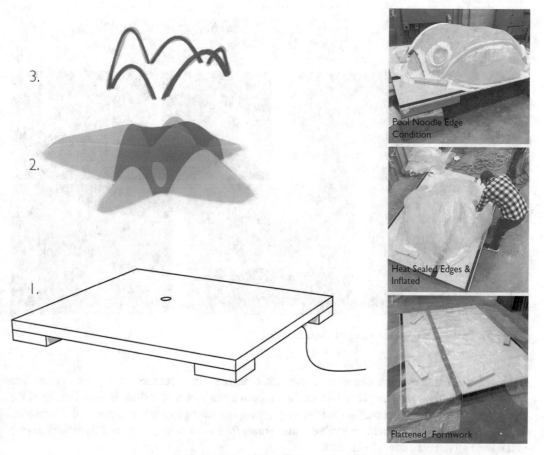

3.

2.

1.

Pool Noodle Edge
Condition

Heat Sealed Edges &
Inflated

Flattened Formwork

Design by Erika Meller, Munenari Hirata, Zachary Price and David Nunez

Fig. 11.37 Exploded axonometric drawing of inflatable formwork

Cardboard Pattern

Flattened Formwork

Fig. 11.38 2D cutting pattern for inflatable formwork

heat sealed seams | inflated | foam arches | concrete poured| plaster | LED lights

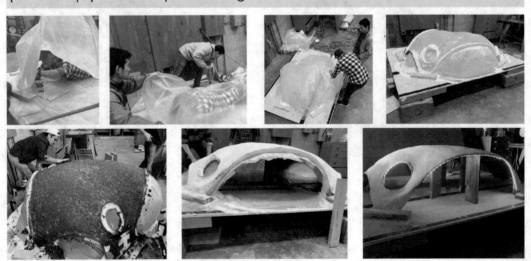

Fig. 11.39 Photos of construction using inflatable formwork

Figure 11.40 is an example of earthen formwork. Even here, laser cut cardboard grids were used to shape and to stabilize the soil. This grid was lifted onto a stack of used pallets to avoid the need for huge amounts of soil. This method seems very promising to explore on a larger scale, where an enormous amount of soil would be needed if the earthen formwork were to continue from the intrados of the shell all the way down to the ground.

Design by James Blanchard, Lillian Cao, Nick Horaney, Hunter Mosier and Cory Peterman

Fig. 11.40 Design ideas for earthen formwork

Figure 11.41 shows the finishing steps of this earthen formwork student project.

Fig. 11.41 Construction sequence for earthen formwork

Figure 11.42 shows an extemely low-tech approach, which simply used sand and soil as the formwork for a fairly large model. The benefit of this is that unskilled labor could be used to create extremely large structures, for almost no material costs at all. In developing countries, this could be a very attractive model. Of course, the disadvantage is an enormous amount of manual labor, which is seen in Fig. 11.42.

Fig. 11.42 Construction
photo for large earthen
formwork

Figure 11.43 shows an overview of several student models. Some of the models used white cement, which is much more expensive than traditional cement, but is incredibly bright in the sunlight.

Fig. 11.43 Various student ultra-thin reinforced concrete shell models

Figure 11.44 shows a large student model of a portion of the design previously described in Fig. 11.23. The students attempted to create a Sugimoto-like image of their model.

Design by Geoffrey Sanhueza, Alex Fernandez, Divyash Agrawal and Zak Collin

Fig. 11.44 Detail of ultra-thin reinforced concrete shell model

The ability to discuss works like Sugimoto's famous photographs of German mathematical plaster models is one of the joys of an interdisciplinary studio. Students spend many hours digging deep into their own personal creations, but they make connections to famous people in history who are directly, or tangentially, associated with shell structures. The line between art and engineering begins to blur when the studio allows students to find that "Art resides even in things with no artistic intentions", as Hiroshi Sugimoto has stated. We hope that the works shown here have convincingly argued that an understanding of the flow of forces in structures from a mathematical point of view can lead to aesthetically elegant structures from a design perspective